MODERN NATURE

MODERN NATURE
Essays in Environmental Communication

LUKE STRONGMAN

Universal-Publishers
Boca Raton

Modern Nature: Essays in Environmental Communication

Universal-Publishers
Boca Raton, Florida • USA
2012

ISBN-10: 1-61233-115-7
ISBN-13: 978-1-61233-115-7

www.universal-publishers.com

Cover image © Irochka | Dreamstime.com

Library of Congress Cataloging-in-Publication Data

Strongman, Luke.
Modern nature : essays in environmental communication / Luke
Strongman.
p. cm.
Includes bibliographical references.
ISBN 978-1-61233-115-7 (pbk. : alk. paper) -- ISBN 1-61233-115-7 (pbk.
: alk. paper)
1. Communication in the environmental sciences. I. Title.
GE25.S76 2012
363.701'4--dc23

2012015922

CONTENTS

I

ENVIRONMENTAL COMMUNICATION

Environmental communication is concerned at a basic level with the way that human beings create their own signs and continually encounter, interpret and use the signs created by others. Environmental communication is the enquiry into the methodologies, theories and practices of the ways in which individuals, organizations, societies and cultures construct, receive, interpret, understand, and use messages about and within the environment by human interactions, thoughts, and informational exchanges using a variety of mediums, tools and technologies. It may include interpersonal and business communication, education for sustainable development, communication within virtual communities, communication in professional contexts such as psychology and medicine, participatory and specialist decision making frameworks, and corporate organizational communication. As Dillman (n.d.) suggests, "[n]o act of communication occurs independently of its environment" (para. 1). A further view of environmental communication is that it is a strategic use of communication processes to support environmental sustainability. As such it is complex and interdisciplinary, and often is seen as useful when there are comprehension gaps between public and expert knowledge.

This book presents ten essays about environmental communication. Chapter one introduces the concept of environmental communication and the ways in which it was conceived, imagined, and developed as a form of interdisciplinary enquiry. Chapter two explores the concept of green communication and education for the sustainable development movement. Chapter three is concerned with one of the major underlying socio-cultural influences of the human/nature divide: that of anthropomorphic or anthropogenic reasoning. Chapter four takes an ecological view of economics and develops an argument for the place of economic intangibles in the modern political economy. Chapters five and six explore specialist aspects of environmental communication practices: Chapter five is concerned with the contexts of psychologist's client and practitioner relationships; and chapter six with the communication domain of the expert courtroom witness. Chapter seven is concerned with exploring the phenomenon of 'social presence' within virtual environments. Chapters eight, nine and ten explore communication practices that are essential within the workplace and organizational environment: Chapter eight frames issues involving understanding ambiguity toleration in business communication; chapter nine explores leadership, management and self-esteem in the organizational communication context; chapter ten discusses the environmental communication contexts of decision-making and organizational trust. The author has written this book for both general and specialist audiences, for students and teachers of environmental communication, and anyone with an interest in the prevalent concerns of 'modern nature' – the current orientation and practices of human communication in natural, virtual and professional spheres. It will also interest students and teachers of workplace organization, including non-governmental organizations and business practitioners.

Varied Characteristics of Environmental Communication

There are many ways that biotic organisms – humans, animals, plants – express themselves in their environment, and in which they exchange and communicate information. This may take the form of an adaptation to environmental conditions (such as noise) or presence, or an absence of light. It may be the way in which the environments affects language and thought (and vice versa) or the development of haptic (touch) communication in virtual reality. It may also include the effects of technology (ergonomics) on human functioning and lifestyle.

Although the focus of this book is on human environmental communication and the corporate topos, we also can learn lessons about environmental communication from the animal world. In order to survive, animals must be able to detect/assess/send signals in their environment, thus animal communication may be affected by environmental conditions at any given time. Consequently, animals evolve signals that distinguish them in effective feedback loops in the environments the signals are used in. For example, Ord, Charles and Hofer (2011) studied the effect of environmental noise on the communication of *amolis* lizards in Puerto Rica and Jamaica, and surmised that:

> Those species communicating to distant receivers and in habitats in which light levels are low (e.g. in full shade) should be under considerable selection pressure to time signals to exploit momentary windows in noise. If so, this might result in predictable variation in signal timing among species living in different habitats and with different ecologies, irrespective of island origin. (p. 55)

Clearly, lizards like people can modify their behavior to communicate in ways that maximize an advantage in their environment. Animals may respond to environmental stimulations as selectively as people do as learned behaviours. Despite the fact that lizards are cold-blooded and people warm-blooded (though sometimes we are prone to draw the zoomorphic comparison), we respond to and learn from our ecologies and constantly communicate what works and what doesn't work well. If genomics has taught us two things, it is that people and animals at the genetic level are not hugely dissimilar, and secondly, that all peoples are genetically very similar. Recent research on the human genome has revealed that "[t]he total amount of DNA carried by an organism [has] little relation to its complexity . . . lilies, salamanders and lungfish all have more DNA than we do" (Finkel, 2012, p. 28). Furthermore, at the biological level our genes are 99.9 identical to one-another. It is how people adapt and respond to their environments and express their genetic inheritance that shapes behavior, identity and communication. As Angrist suggests, "[w]e are more than our genes – we're the expression of our genes. Looking at differences in expression is going to be much more informative than genotype." (2010, p. 131) Humans then are characterized by both adaptability and expression. Both of these qualities account for the breadth and depth of human interaction with their ecologies, societies, and workplaces – with environmental communication.

As Broditsky (2011) relates, a dominant feature of human intelligence is adaptability – the capacity to invent and rearrange conceptions to suit the changing demands of the environment (p. 43). Languages reflect both this capacity and the diversity of its effects. In this way, language molds the individual into the environment. Broditsky regards it as an "invaluable guidebook developed and honed by our ancestors" (p. 45). Differences in the environment create differences in language and thinking, and vice versa.

The interaction and effects of people, their environment, and technology is complex. As Germonprez and Zigurs (2009) suggest, "one perspective argues that individuals continually acquire knowledge for action in a dynamic, changing environment and they alter their actions, language, and technology in response to that dynamism" (p. 23). In this respect, people use and adapt technology with changing knowledge and capability. This technology use may reflect the two characteristics of communication: 1) that it can be deterministic in nature – reinforcing social and structural norms, and 2) that it is emergent – capable of creating new meanings (p. 23).

One recent adaptation to virtual technologies is the incorporation of haptic 'touch' feedback devices. These are used for example in medical devices for minimal invasive surgery, in remote manipulation by robots in space, and also in collaborative working environments (CVEs). These are digital spaces that allow remote users to work together and to experience tactile communication exchanges in doing so. However, collaboration also requires a shared mental representation. This is referred to as a frame of reference or common ground. As Chellali, Dumas and Milleville-Pennel (2011) suggest, haptic feedback is a direct form of human interaction; it can be used to express feelings of closeness, intimacy and trust (p. 318).

According to Rasmussen (1983) there are three categories of human behavior: skills, rules, and knowledge. The sensory-motor performances are based at the skills level, shared by a reflex behavior. Haptic communication requires physical contact in this modality and is possible even in remote communication, but may involve the oldest regions of the brain.

Environmental Communication and the Concept of Place

The experience of place in environmental communication is a dimension formed by people's relationship with physical settings and activities. As Najafi and Shariff (2011) points out:

> Place identity and 'sense of place' are sometimes used synonymously and describe the quality of peoples' place rela-

tionships. However the two terms differ in subtle ways. Firstly, in studying human-place bonding and attachment, while sense of place may refer to the assignation of value and bonding to a place in terms of an overarching impression. (p. 187)

Sense of place in environmental communication is important to the quality of the environment and human integrity.

Furthermore, place can be distinguished from space in as much as it represents an affective bond between a person and a particular setting. Place is thus concerned with human values whereas 'space' may be a natural term. However, places may be in space and have a unique character. Consequently, sense of place helps to maintain the quality of the environment. It may be involved in connecting people to shared experience and it may help to consolidate community ties. However, sometimes places have a lack of identity or no 'placedness'. As such, they do not convey the minimum of feelings and meanings to people and may be indistinguishable from any other similar environment. In such space one can usually still discern basic features such as perspective, but no implicit feeling of relatedness or relatability. Although designers, architects, and landscapers may create a sense of place where none existed before, some spaces featuring 'placedness' occur naturally. The added element which changes environment to place is the experience provided by the phenomenology of perception. This 'being in the world' is related to placedness, and is a fundamental quality of psychological existence. It may be a product of experiences of quality in existential space. Najafi and Shariff (2011) explain how the phenomenologists' concept of Topophilia is used to represent 'character of place' and 'spirit of place' (p. 188).

As Najafi and Shariff (2011) explain, the experience of place may have several levels (p. 188). The first is familiarity, which may or may not be associated with depth of feeling and meaning. The second level is ordinary familiarity, perhaps the unconscious perception of place. This level is experienced more on a collective and cultural level than a personal level, signifying a deep attachment to place and close attention to its symbols. The third level is that of 'profound familiarity' involving the 'existential insideness' of a person and integration with place (Najafi & Shariff, 2011, p. 188).

The communication of relatedness to place and sense of place is important to environmental psychologists as the physical environment plays a role in both the immediate and long term impact of human behavior and mental health. Consequently, the design of place involves both emotional and functional qualities. Thus, 'sense of place' encompasses the relation between humans and place. Najafi and Shariff (2011) suggest it is defined by three elements: location, landscape, and personal involvement. Furthermore, Rogen et al. (2005) define a psychologically comfortable environment as being legible, capable of qualitative perception, and preferable, in other words, compatible for human living (Cited in Najafi et al., 2011,p. 189).

Najafi and Sharrif (2011) identify seven levels of human interaction with 'sense of place' in the environment (p. 188):

- Not having any sense of place
- Knowledge of being located in a place
- Identifying with a place and its purposes
- Involvement in a place
- Belonging to a place
- Attachment to a place
- Sacrifice for a place

Involvement with sense of place as a phenomenological experience is thus psychological, social/interactional, and physical. Peoples' relationship with place may also be transactional, i.e., people either give or take, or experience positive or negative associations with place. There may be eleven factors influencing place attachment: 1) the emotions people feel for place, 2) socio-demographic characteristics, 3) environmental experiences, 4) the nature and type of peoples' involvement with place, 5) peoples' degree of familiarity with place, 6) peoples' expertise with place, 7) the experience of culture in relation to place, 8) peoples' satisfaction with place, 9) peoples' preference or attachment to place, 10) the kinds of activities people experience with place, and 11) the phenomenological experience of place itself (Najafi & Shariff, 2011, p. 191).

Environmental Communication and the Human Organism

Environmental communication spans both the material and nonmaterial in human interaction. Behavior and thought, as Simpson (1964) suggests, "living things have been affected for . . . billions of years by historical processes . . . results for those processes are systems different in kind from any nonliving systems and almost incomparably more complicated" (p. 106). This suggests that the laws of perception, behavior and communication are ecological and defined in terms of the organism and its environment (Gibson, 1986). As Stewart (2007) suggests, "individual psychologist believe that . . . lifestyle depends upon myriad organismic and environmental influences that people experience as they pursue the tasks of family, work and community" (p. 67). Adler (1938), for example maintained that organismic and environmental variables are inherently interactive and are significant in affecting personality. The relationship between a conscious organism and its function are complex; the actions of the organisms are intrinsically to do with its functioning and 'construction'. For German enlightenment philosopher Immanuel Kant (1784 version), an organism can be distinguished from a machine for this reason: "The parts act together to meet the thing's purpose; their actions, however, have nothing to do with the thing's construction. The thing and its parts rely on efficient causes arising from outside themselves for their origin and function" (p. 46).

Consequently, a conscientious organism that is capable of self-organization and holds an ecological property are synonymous; the parts are not easily separated from outside of the whole they comprise. Such and organism is not fractionable (Petrusz & Turvey, 2010, p. 47). For example, in the self-organizing process in the evolution of a bird's wing, the propulsion and lift it offers the bird are not defined outside of the form and function of their whole (Petrusz & Turvey, 2010). As Barad's concept of 'agential realism' (as cited in Francovich, 2010, p. 311) "captures the paradox of the observer and the observed by means of the argument that relationship precedes related and that we as individuals/organisms are in a flux of continuous emergence that is fundamentally nondual". Environmental communication takes a slightly materialistic slant towards the investigation of the form and function of the human organism and the environment in which that human is situated – which it co-creates. After all, living organisms are subject to the same physical laws as any other entity; they are nevertheless fewer in number. This has led to the belief that the conscious organism is divisible into theories of the non-living forces. However, this deconstructivist tendency runs against the properties of an ecological system: that it comprises the environment which defines its purpose as living entity. Living systems are creative.

Petrusz and Turvey (2010) argue that there are four kinds of laws that define a 'science of the animate' in psychological and environmental communication terms. These are as follows: Firstly, human behavior is guided by the capacities of the organism-environment as a whole, defined in terms of intentions or opportunities for realizing them. They may be understood in terms of lower level physical features and non-fractionable higher level features. Secondly, human behavior is guided at the ecological scale of the organism and the environment not of their components. They are defined by a meaningful physics of semantic affordance. Thirdly, the human organism is more general than (not a special case of) the laws of an animate system. It is not the case that only laws that govern animate systems are those that govern their physical components (p. 65). As Petrusz and Turvey suggest, "the causal entailment of animate systems exceeds that of physical (inanimate systems) (p. 64). Fourthly, they make reference to themselves in an impredicative manner, perception-action is self-referential.

From a perspective of environmental psychology, the study of human-environment relationships is a wide field and includes topics such as: territoriality, personal space, crowd effects, environmental stress, scholarly, business and work environments, home environments, environmental influences on behavior, attachment to place, isolation and contained environments, the assessment of environments, and it also values beliefs and attitudes concerning the environment (Stewart, 2007, pp. 68-69). Thus the literature of environmental communication and psychology is informed by laboratory, virtual, field, and ecological settings as well as from architecture, design, and industry. There are three main paradigms. Firstly, the adjustment paradigm views the environment as physical, social, interpersonal, all of which are communicative influences on people as they adapt, respond and function in the environment, for example to stress, arousal or stimulus (Stewart, 2007, p. 69). The opportunity structures paradigm studies how people function in a socio-physical environment to meet their needs, goals, wants, and how they perform their roles within it. Thirdly, the socio-cultural paradigm is the study of how culture and society shape the socio-physical environment and relationships within it (Stewart, 2007, p. 69). Consequently, environment may shape behavior and to some extent personality, but it is also true that peoples' attitude and behavior towards the environment can affect the psychological well-being of the individual and that of other people (Stewart, 2007, p. 78). Obviously, having a tolerant, open and enhancing attitude to the environment will foster a more healthy psychological state of being as the environment will be co-created in a more accepting and balanced way and be more likely to be sustaining.

As Binne-Dawson suggests, (1982, p. 397) the bio-social approach to environmental psychology and environmental stress is permissive of three main orientations towards the environment: 1) Behavior occurs within an environmental context which imposes condition and constraints on the functions of the individual within it; 2) Some environmental conditions such as over or under stimulation or climate extremes, will have more general effects on systems of response within the individual; and 3) Behavior is instigated by particular environmental attributes and characteristics. Environmental adaptation is a social as well as physical process. Social systems as well as physical systems lead to adaptations in perceptual and cognitive abilities, attitudes, and values, which in turn affects ecological survival (Binnie-Dawson, 1982, p. 199). As Lakoff (2011) suggests, people are nature – "nature is not separate from us . . . what is good is the use of nature that doesn't use up nature" (para. 5).

Environmental Influences on Social Behavior

As Stewart (2007) suggests, "'Behavioral settings refer to systems emerging from the temporal and spatial arrangement of social and physical characteristics of small-scale environments" (p. 71). One of the main environmental influences on behavior is crowding. There is some evidence that crowding has a negative impact on the social behavior of children. The reasons for this are that it may produce an adverse subjective experience on more vulnerable and younger children leading to relationship difficulties arising from restrictions on behavioral freedom, excessive stimulation, or intimacy that is too intense or inappropriate (Stewart, 2007, p. 70). Under-population of the environment may also lead to less than optimum development of meaningful interpersonal roles. Under-population theory holds that there will be a maximum amount of positive psychological benefits when there are more behavioral settings and roles to be fulfilled and experienced than there are individual people competing to assume those roles (Stewart, 2007, p. 71). Under-populated social environments according to Barker (1968) and Schoggen (1989) operate in a *centripetal* manner such that interpersonal communications and behaviors provide feedback, participation, and cohesiveness to maintain a frame of social reference for any given task or activity (Stewart, 2007, p. 72). Overpopulated social environments may lead to the marginalization through *centrifugal* social forces.

Gibson (1979) is attributed with founding the field of ecological psychology with influences on both cognitive psychology and environmental communication. Ecological psychologists characterize 'information' as the 'patterning of energy that occurs as it passes through a medium and interacts with objects" (Jordan, 2009, p. 130). There is a quite profound difference between this model of human perception and that of cognitive psychologists. Instead of emphasizing the ambiguity of information and the brain's capacity for specialization in processing it, ecological psychologists distinguish less between the person and their environment, arguing instead that the brain and sensory system 'resonate to environmental information' (Jordan, 2009, p. 130). Under this conceptualization, the behavioral possibilities that a stimulus affords is equally as important as its perceptual qualities. Thus instead of emphasizing the ability to perceive and cognate an object internally (to assess size, shape and color), ecological psychologists emphasize the affordances of such perception in terms of the possibilities it offers the individual for use – what can be 'done' with an object.

As Jordan (2009) points out, the ecological psychology view of human-object relations affordance is consistent with the theory of perception-action coupling termed 'Theory of Event' coding. Theory of event coding asserts that: 1) Actions are planned in terms of the distal effects produced, (i.e. outcome in the environment), and 2) cognate perception, planning, and action use overlapping neural resources (Jordan, 2009, p. 131). This would imply that perception is inherently intentional because planning alters the neural configurations of perception, which in turn is consistent with the ecological psychology view that perception is made in terms of affordances (behavioral possibilities detected by perception (2009, p. 131). Affordant properties are emergent contextual and based on perceptual perspectives, as Francovich (2010) suggests, " . . . the behavior of any organism can be understood as melded or continuously fused to the medium, surfaces, and substances of contact experience . . . in continual emergence" (p. 313). Intentional induction, for example putting one's hand over one's eyes to scan the horizon is one such example. Ecological psychology also renders the explanation of experiencing the intentions, desires or beliefs of another in less fractal terms than computational approaches. We experience the mental contents of another person because we experience the distal effects they generate in the environment, as well as the actions and means that generate in the context of our own embodied action-effect contingencies. This differs from either the computational or cognitive models of mind in so much as, "instead of being 'trapped' inside non-observable, symbol-manipulation systems, as is assumed in computational approaches, the resonance-based approach asserts intentionality as entailed in the continuous synergistic, multi-scale couplings that constitute intentional context, including neutrally embodied action-effect regularities, neuro-muscular architectures, and the external regularities such architectures give rise to and are embedded within" (Jordan, 2007, p. 137). This has large repercussions for environmental communication in so far as the ability to couple individual intentional contexts within group contexts is fundamental to an ability to generate, use, and sustain sign systems (Jordan, 2007, p. 139). As Craik (1977) suggests, the perceptions an observer derives from a setting depend not only on physical attributes, but other factors such as: 1) the cognitive set, 2) how the place is encountered, 3) the available formats for responses and affordances (p. 149). Furthermore, as Francovich (2010) contends, it is inherently related to the creation of self and environment, "self [is] a dynamic emergent process that comprises the 'living present' . . . whereby reflective consciousness creates the symbolic world of spatial and temporal independence and thereby stabilizes the appearance of enduring subjects before static and/or dynamic objects (artifacts of reflective consciousness) in a timeless space regulated or metered by spaceless time" (p. 315). However, environmental perception is arguably at its most complex when the encounter is with other people. As Adams (2007, p. 24) states, interrelating is fundamental to human culture and can be characterized in three ways: Firstly, it is there from the beginning of every presencing moment. Secondly, it forms the basis of human interaction – it is the progenitor of health, justice and compassion for humankind. Thirdly, it is the 'ever-present' path by means of which we carry out our 'inter-existence'. Fourthly, there is an ethical imperative to cultivate interrelation in a way which serves others and the non-human natural community. Furthermore, arguably the socialization of inter-relation contributes to the development of human consciousness deriving from environmental communication. As Francovich (2010) suggests:

The significant gestures (Mead's term) that organisms employ are happening on an evolutionary ladder of increasing coordination via communication amongst entrained interlocutors that eventually, and through pressures of increased social complexity and evolved neurobiological structures, result in a 'phase shift' in communications. The conversation of gestures, according to Mead, results in homo sapiens internalizing via the significant symbol first specific to others (family) and then through a progressive process of role taking and role playing eventually come to the generalized other and creation of the reflective self. This internalised [sic] generalised [sic] other becomes the 'me' of the reflective consciousness and is the beginning of the deep patterning that we come to know as our subjective selves. (p. 314)

Consequently the need to improve the ecological environment is not only a biological imperative but a social, technological and ethical one, requiring the exercise of authentic relationships. Inherently our use of language links us to the perceptual participation in the phenomenal world in a system of inter-relations, signifying the evolution of the human species, the individual development of the child and also exists in the moment of co-presence with others (Adams, 2007, p. 50). However, as Adams suggests, extreme versions of linguistic primacy may lead to the anthropogenic argument that humankind is both self-sufficient and self-sustaining (Adams, 2007, p. 51). Thus human communication and human culture are interrelated. They both have pattern, some form of order, and consequently an evaluative dimension that is dynamic over time.

References

Adams, W. A. (2007). 'The primacy of interrelating: Practicing ecological psychology with Buber, Levinas and Merleau-Ponty'. *Journal of Phenomenological Psychology, 38,* 24-61.

Adler, A. (1938). *Social interest: A challenge to mankind.* (J. Linton & R. Vaughan Trans.). London: Faber and Faber Ltd.

Angrist, M. (2010). *Here Is a Human Being. At the Dawn of Genomics.* New York: Harper Collins.

Barker, R. (1968). *Ecological psychology.* Stanford, CA: Stanford University Press.

Binnie-Dawson, J. L. M. (1982). 'A bio-social approach to environmental psychology and problems of stress'. *International Journal of Psychology*, *17*, 397-435.

Broditsky, L. (2011). 'How language shapes thought'. *Scientific American*, *304* (2). Available from http://www.scientificamerican.com/article.cfm?id=how-language-shapes-thought, 42-45.

Chellali, A., Dumas, C., Milleville-Pennel, I. (2011). 'Influences of haptic communication on a shred manual task'. *Interacting with Computers*, *23*, 317-328.

Craik, K. H. (1977). 'Multiple scientific paradigms in environmental psychology'. *International Journal of Psychology*, *12* (2), 1467-157.

Dillman, R. (n.d.) 'The Communication Environment'. *HFCL Tutorial*. Retrieved from: http://www.rdillman.com/HFCL/TUTOR/ComEnv/ComEnv1.html

Finkel, E. (2012). *The Genome Generation*. Melbourne: Melbourne University Press.

Francovich, C. (2010). 'An interpretation of the continuous adaptation of the self/environment process'. *The International Journal of Interdisciplinary Social Sciences*, *5* (3), 307-322.

Germonprez, M., & Zigurs, I. (2009). 'Task technology, and tailoring in communicative action: An in-depth analysis of group communication'. *Information and Organization*, *19*, 22-46.

Gibson, J. J. (1986). *The ecological approach to visual perception*. Hillsdale, NJ: Erlbaum.

Jordan, S. (2009) 'Forward-Looking aspects of perception-action coupling as a basis for embodied communication'. *Discourse Processes*, *46*, 127-44.

Kant, I. (1784). *What is enlightenment?* Retrieved from http://www.english.upenn.edu/~mgamer/Etexts/kant.html

Lakoff, G. (2011). 'Why environmental understanding or "framing" matters: An evaluation of the EcoAmerica Summary report'. Retrieved from: http://not4p.wordpress.com/resources-and-readings/why-environmental-understanding-or-framing-matters-an-evaluation-of-the-ecoamerica-summary-report/

Najafi, M., Shariff, M. K. B. M. (2011). 'The concept of place and sense of place in architectural studies'. *International Journal of Human and Social Sciences*, *6* (3), 187-193.

Ord, T. J., Charles, G. K. M. Hofer, R. K. (2011). 'The evolution of alternative adaptive strategies for effective communication in noisy environments'. *The American Naturalist*, *177* (1), 54-64.

Petrusz, S. C., Turvey, M. T. (2010). 'On the distinctive features of ecological laws'. *Ecological Psychology, 22*, 44-68.

Stewart, A. E. (2007). 'Individual psychology and environmental psychology'. *The Journal of Individual Psychology, 63* (1), 67-85.

Rasmussen, J. (1983). 'Skills, rules, knowledge: signals, signs and symbols and other distinctions in human performance models'. *IEEE Transactions on SMC, 13* (3), 257-267.

Rogan, R, O'Connorb, M., Horwitza, P. (2005). 'Nowhere to hide: Awareness and perceptions of environmental change, and their influence on relationships with place.' *Journal of Environmental Psychology, 25*, 147-158.

Schoggen, P. (1989). *Behavior settings: A revision and extension of Roger G. Barker's Ecological Psychology.* Stanford, CA: Stanford University Press.

Simpson, G. G. (1964). *This view of life.* New York: Harcourt, Brace & World.

2

GREEN COMMUNICATION
Education for Sustainable
Environmental Development

This chapter draws on a range of literature in the field of sustainability to outline the main ways in which sustainability has been defined. It explores the central debates within the emergent sustainability movement and traces the interdisciplinary connections between sustainability values in indigenous models, eco-psychology, business, and higher education. It also seeks to synthesize the core principles of sustainable practices for individuals, business and educators.

Introduction

According to Calder and Clugston (2003), only after the 1992 Rio Earth Summit did the term 'education for sustainable development' (also known as EFS, or 'education for sustainability') enter the vocabulary of educational reformers (p. 10003). They point out that issues of sustainability were first accepted as areas of study in higher education through the influence of non-governmental organizations, businesses, and environmental lobbyists. After the Brundtland report of 1987 by the World Commission on Environment and Development (WCED), (as cited in Calder & Clugston, 2003, p. 1) government support in the United States, Europe, and some developing countries such as Chile, Ghana and Denmark helped to bring sustainability issues to the attention of academic disciplines and the professions. However, by the second decade of the new millennium issues of climate change and sustainable development are at the forefront of international environmental and economic discussion. The Rio + 20 Earth Summit in 2012 will highlight the tension between economic development which is environmentally destructive and that which envisions possibilities for a more sustainable and future-focused ecological inhabitance and the overriding need for co-operation as Earth's population approaches seven billion.

In the new millennium, sustainability is a critical topic on the political agenda internationally. At the 2009 Copenhagen Climate Conference (COP15 United Nations Climate Change Conference 2009), the New Zealand government announced a commitment of $45 million to fund a research network to reduce carbon emissions from farming. The Copenhagen Global Research Alliance will investigate the relationship between agricultural output and greenhouse gas emissions. The New Zealand-led initiative involves a consortium of 20 countries (including The United States of America, Canada, India and Australia) that plan to investigate farming practices and explore the development of new technologies which reduce emissions from livestock, cropping and rice production. Such developments may assist in the mitigation of climate change and in adaptation to 'green' technologies.

Interest in sustainability as a global movement has grown from the realization that the world has finite resources, which people may be consuming more quickly than they can replace, discover, or invent. Goldman et al. (1999) state:

> Among the major factors contributing to the degradation of the environment are population pressures, particularly widespread poverty. (*Educating for a Sustainable Future*, UNESCO) From 2.5 billion in 1950, world population is projected to reach more than eight billion by 2025. The human population places the greatest stress on Earth's resources and natural processes. The US Geological Survey (USGS) estimates that the use of air, water, and other natural resources has increased by a factor of 10 in the past 200 years. A cycle of consumption and overuse is perpetuated as areas are developed, resources exhausted, and populations relocate. Excessive fishing, harvesting, and grazing result from increased demand for food, goods, and services, which increases the demand for natural resources and land use. (p. 15)

Although Daly's (1996, p. 5) model of steady state systems holds that the flow of energy in the universe is constant and the increase of entropy in the overall system is negligible, no theories that are based solely on the physical properties of the universe can easily be applied in delivering sustainable societies. O'Sullivan and Painter (2006) point out that in Daly's view, sustainable growth is not possible since the economy is an open-ended system of the earth's ecosystem. This ecosystem itself is finite and materially 'sealed off' from the universe, except for the sunlight the earth captures, the heat it reflects into space, and the gravitational effects exerted in it by other bodies (O'Sullivan & Painter, 2006, p. 1). As Hamilton (2010) points out, one of the core issues for environmental sustainability is the human-made changes to the naturally occurring carbon-cycle brought about by fossil-fueled industrial energy consumption:

> For nearly three million years the natural carbon cycle has ensured the atmosphere has contained less than 300 parts per million (ppm) of Co_2, just the right amount to keep the planet at a temperature suited to the flourishing of a rich variety of life. But human industrial activity over the last two to three centuries has disturbed this balance. When we dig up and burn coal, over half of the Co_2 released is absorbed by land and ocean sinks. The rest stays in the atmosphere, some of it for a very long time. A quarter will still be affecting the climate after a thousand years and around 10 per cent after a hundred thousand years. (pp. 8-9).

This effect may be longer than the life of radioactive material from nuclear waste (Archer, 2009, p. 1). However, O'Sullivan and Painter (2006) consider that sustainable development may be possible by sustaining natural capital through regimes implemented to maintain the biotic economy (pp. 1–2). As Wilson (1998, p. 277) puts it, "We [*Homo sapiens*] are the first species to become a geophysical force, altering Earth's climate, a role previously reserved for tectonics, sun flares, and glacial cycles." (Wilson's comment is debatable; in terms of biomass the human population is vastly exceeded by the insect population). Such macroscopic aggregations of the effect of *Homo sapiens* on the environment may be an inspiration for many, but to achieve sustainability, the many have to act in an attempt to change their assumptions and behaviors.

As early as 1934, the architect and writer Lewis Mumford in *Technics and Civilisation*, (as cited in Watson, 2000, p. 288) had proposed that technology was driven by capitalism. He posited three stages in its evolution: the *eotechnic* era characterized by engines made of wood and driven by wind or water power, the *palaetechnic* era, which corresponds to the first industrial era brought about by the use of iron and steam-powered engines, and the *neotechnic* era, characterized by the use of synthetics, alloys and electricity. In the current era (or second industrial revolution, sometimes described as the post-industrial era despite its reliance on fossil-based fuels), alloys and synthetics are the most consumptive of carbon-based resources and the most generative of non-biodegradable waste in the form of greenhouse gases. Writing on the cusp of the atomic era, Mumford's contribution to the sustainability debate arose from his belief that the social and economic organization of society was not necessarily aligned with its uses of technology, leading to environmental damage.

Sixty years after Mumford, the Stern report (2006) considered that "most climate models show that a doubling of pre-industrial levels of greenhouse gases is very likely to commit the Earth to a rise of between 2–5°C in global mean temperatures" (p. 3). It also suggests that a rise of 5°C in global mean temperatures in the period between 2030 and 2060 would fall outside of the predicted range of historical tolerances for maintaining the status quo of human habitation on the Earth. The Stern report is a stark reminder that issues of sustainability require understanding, forethought and proactive engagement on a very wide scale.

To begin with, sustainability needs critical definition. 'Weak' sustainability holds that some substitution between naturally occurring and synthetic resources is tolerable, as long as the welfare of people within the ecosystem does not decline, thus, 'optimality' is preserved (Beckerman, 1994, p. 195). Daly's (1996) view is that 'strong' sustainability requires synthetic and natural capital to be developed separately (p. 10). Thus, there is trade-off between utility and idealism in these two definitions.

The study by Sustainable Aotearoa New Zealand (2009) suggests that the strong sustainability model contains a concentric arrangement of the biosphere (network of biotic relations), the sociosphere (network of social relations) and the econosphere (network of economic relations). The study argues against the scientific viability of what it terms the 'triple bottom line' cultural model (Adams et al., 2009, p. 6). However, it overlooks the psychological and architectural topos of urban infrastructure (the situation in which most people around the globe live), where sustainable action (and practical difference) occurs at the intersection of the sociosphere, the econosphere and the biosphere.

The model this study offers is precisely the model that industry has followed for decades – one that places the economy as the central hemisphere of a sociosphere within a larger biosphere, which, although not wholly characterized by sustainable action, is nevertheless not fundamentally antithetical to it, unless one regards human-made infrastructure alone as the basis of the real economy. The point is that the econosphere and the sociosphere are consuming the biosphere. In the sustainability paradigm we need to draw distinctions between seven forms of capital:

1. natural capital (aquifers, biomass, soil, microbiology, and atmosphere)
2. physical capital (equipment and infrastructure)
3. human capital (embodied skills, services, and biomedicine)
4. intellectual capital (information, disembodied skills, and knowledge)
5. social capital (formal/informal relations among workers, and organizations)
6. financial capital (savings, loans, sale of stocks, and sale of bonds)
7. cultural capital (art and design, qualifications, standards, and semiology)

Strong sustainability is concerned with restoring a balance between all renewable and non-renewable forms of capital (forms one to four), which may require the application of Bourdieu's (1986) forms five to seven. O'Sullivan and Painter (2006) point out that, while most international assessments have concluded that resource depletion is endemic and waste assimilation pandemic, in some areas of society there is little concern at the individual level (p. 2). This lack of interest leads to inadequate provision of programs for education for sustainability in the environmental politics of many Western economies. In New Zealand for example, while engineers have long recognized a responsibility to maintain the biosphere, sustainability issues have latterly been recognized only by the Resource Management Act 1991 (RAM) and in policy statements such as the Sustainable Development for New Zealand Programme of Action (Ministry for the Environment, 2003), the Sustainable Water Programme of

Action (Ministry for the Environment, 2009) and the work of Sustainable Aotearoa New Zealand (SANZ). One proposed solution of 'carbon taxes' on the processes of carbon emissions is heavily debated internationally and currently seems difficult to implement at a national level due to the multi-lateral interests of the global economy. It would only work if the revenues are consumed by increasing spending on the development of 'green' products or on direct environmental and ecologically sustainable goods and services.

To do the concept of sustainability justice in a post-Brundtland pluralistic world, insight into both sustainability issues and strategies for remedying them is required from a variety of disciplines, perspectives and methods. We need to examine more closely the interaction between societal structures and human behaviors, as well as any new possibilities for demographical organization, systems of socialization, economic production, and ordered patterns of consumption in the environment that address sustainability issues. Sustainability as an interdisciplinary field includes the following disciplines: economics, engineering, humanities, anthropology, sociology, psychology, architecture, geography, agriculture, planning, and legal systems.

Dr. Kate Hewston (personal communication, February 2, 2009) states that sustainability involves the systematization of the relationship between reductionism and holism with regard to social, economic, and environmental constraints. As such, there is a need – as O'Sullivan and Painter (2006) explain – to 'move away from the "command and control" paradigm', not towards a streamlined, status quo, 'more market' approach, but towards individual and collective recognitions of people as participants in a 'self-organizing' system (p. 13).

Ecological systems are characterized by energy flow, interdependence, adaptation, cycles, and diversity. In any 'strongly' sustainable system there is a free-flow in which relationships between constituent parts are nurtured to produce a steady-state economy in which the aggregate throughput of matter and energy over time is constant. Concomitant with this is the realization that individuals and populations do not live apart from nature but within it, in symbiotic and other relationships that all use energy and biotic materials, forming an ecosystem. As far as education is concerned, sustainability is usually associated with three concepts – environment, economics, and the concept of society. The following considerations may apply to any syllabus for sustainability education:

- green policy and practice
- education for change – awareness and behavior
- communication and planning
- business efficiencies and distribution of resources
- time management, management of resources
- human needs and human rights
- law and governance
- local knowledge and scientific knowledge
- interdisciplinary subject content
- 'green' awareness in pedagogy
- links between the community, the professional sector, and industry
- the need to make ethically positive, active choices about sustainable practices

Thus definitions of sustainability are, in principle, broad and encompass these fundamental dimensions. At the core of the sustainability movement, within its strategies for preserving a life-sustaining biosphere is human development.

Definitions of Sustainability and Policy Positions

The Brundtland report (World Commission on Environment and Development [WCED], 1987) describes sustainability as "meeting the needs of the present without compromising the ability of future generations to meet their needs" (p. 1). Defined in terms of economics, sustainability is a measure of "non-declining per capita utility between inputs of three resource forms: labour [sic], capital, and natural resources" (Moffatt, Hanley, & Wilson, 2001, p. 75).

Within the politics of sustainability there are tensions between the supporters of sustainable transformations and those who see no need to act, or are satisfied with the status quo. These views may be expressed as a continuum with individual freedoms at one end and governmental control at the other. From an economist's perspective, natural and man-made capitals are substitutes; they are frequently not differentiated in economic terms, and neo-liberals may believe that environmental problems can be overcome by investing financial capital in innovative technology. Technological sustainability allows development in the form of qualitative improvement, such as new technical solutions or changes to environmental problems caused by the application of new knowledge. To return to an economic focus and use a metaphor from Alan Greenspan (2007), sustainability may involve the 'creative destruction' of the old model of environmental depletion, and its replacement in a naturally occurring generative sense by the new model of sustainable practices. Isn't this precisely what nature does, by working in cycles?

As a counterpoint to theories of economic sustainability, Beck (1982, p. 6) identifies the so-called radical 'red-green' literature (the two colors represent frequencies of energy consumption), which holds that capitalism systematically undermines environmental sustainability. Ecological disruption is an inevitable consequence of the way in which society is organized under the profit motive of capitalism (if not neo-liberalism itself). This is also the fallback position of the environmentalist movements whose advocates claim that human society creates a 'metabolic rift' in the environment in which systems are overloaded and cannot cope with waste generation. A treadmill is created where businesses are constantly required to expand to make profits, (and government subsidies in the form of tax breaks support business expansion), while employees are coerced into political agreement to keep their jobs. Some of the resultant capitalist gain comes at the expense of the environment.

At the macro level, Wilson claims that the current loss of biodiversity is the greatest since the end of the Mesozoic era, some sixty-five million years ago (Wilson, 1998, p. 294). It is possible that the environmental damage caused in the twentieth century to the diversity of non-human biota will require millions of years to restore. Historians of the twenty-second century may look back on the current era with dismay (if they are able to).

Similarly, others claim that objectivist science has been harnessed to meet the vested interests of business groups. The Danish environmentalist Bjorn Lomborg (2001) has thrown doubt on the extent to which human civilization has influenced climate change, forcing others to ask what constitutes proof of climate change and whether science is itself a construct that may be used for political ends. Others, such as the former American Vice-President Al Gore, argue that these skeptical counter-beliefs distract attention from inconvenient truths about the consequences of the impact of human society on climate change (Gore, 2006) and dissolve focus from more practical environmentally sustainable solutions. As Barrett states, ". . . the members of each generation are trustees of the natural world for the benefit of all people – those who are alive and those not yet living. As trustees, we must strive to understand the mechanisms and interactions of the ecosphere and how our actions impact upon them" (2010, p. 1-2).

Sustainable development requires that the rate of depletion of non-renewable resources should be restricted to a level whereby their use should foreclose as few, and keep open as many, options for the future as are possible (WCED, 1987, p. 3). Strong, sustainable development rejects the idea that natural capital and man-made capital are interchangeable. The preferred option is an acceptable balance between the production and consumption patterns of human society in natural ecosystems. Unsustainable development is founded on the attitude, often held in ignorance, that the world and its resources are a giant open system and that it is a human 'right' to deplete natural resources while polluting the environment and producing goods that end up as waste.

However, as Karl-Henrik (2002) points out, natural cycles surround society and comprise the parameters of life in whatever form (p. 61). Sustainability is dependent on natural flows from nature's production, the biosphere, and materials from the Earth's crust. If everything is made of atoms that were dispersed 3.5 billion years ago and have organized themselves into concentrated structures that form the biota of the biosphere, the aim of sustainability is to treat this natural flow as a form of 'interest rate' from nature, rather than as a toll on its underlying capital (Karl-Henrik, 2002, p. 61). Karl-Henrik's concept of 'interest rate ecological sustainability' is similar to Eckersely's notion of 'free market environmentalism' and McAfee's imperative that 'nature be sold to save it' (as cited in Clark, Massey, & Sarre, 2006, p. 121).

However, the Brundtland report (WCED, 1987) holds that the neo-classical economy model applied by the main industrial powers is inexorably unsustainable and detrimental to the long-term health of the planet. The report identifies the over-use of environmental resources as the cause, suggesting that economics and ecology bind us in ever-tightening networks, necessitating increased cooperation and policies that sustain and expand the earth's resource base. The Brundtland report identifies limitations that our current state of technology and social organization impose on the ability of the environment to meet present and future needs. In this context, the term 'ecological footprint' refers to the amount of productive land required to support society with available technology. It is expressed by the PAT formula: Population size multiplied by per capita affluence (consumption), multiplied by the application of technology used in sustaining consumption (Wilson, 1998, p. 282). Ecological footprint is different from the carbon footprint, which is a measure of fossil fuel consumption per capita, offset by the available carbon sink from photosynthesis in the ecosystem. The 'food fork' is the percentage of arable productive land divided by the rate of population growth.

Within the sustainability paradigm, three separate but interrelated movements take significance. The first of these is the Gaia hypothesis, formulated by James Lovelock in the mid-1970s. Founded on the holistic principle that Earth is a living entity and that separation between humanity and the environment is arbitrary and anthropomorphically biased, Lovelock's hypothesis conceives "Gaia as a complex entity involving the earth's biosphere, atmosphere, oceans and soil; the totality constituting a feedback or cybernetic system which seeks an optimal physical and chemical environment for life on this planet" (Lovelock, 2000, p. 10). As such, he views the earth's ecosystem as a 'living equilibrium' governed by homeostasis (p. 10). Affirming the Gaia hypothesis means recognizing that human relations are intrinsically connected to the natural world.

The second interrelated movement is the 'deep ecology' movement. It views humanity as arising from and ultimately only compatible with the tolerances of the natural ecosystem. The concept of deep ecology originated with the Norwegian philosopher Arne Naess, who posited seven principles of grass-roots ecocentricism. These principles are:

1. *Rejection of the human-in-environment image in favor of the relational, total-field image:* This principle holds that it is only by examining humanity in relation to the totality of life forms and life systems that anthropocentric behavior may be corrected.

2. *Biospheric egalitarianism:* According to this principle, respect for all forms of life on earth is a basic tenet and value.

3. *Diversity* and *symbiosis:* This involves a broad-spectrum approach to the potential for life, the preservation of genetic culture, and the plenitude of life forms.

4. *Anti-class posture:* This principle seeks to neutralize group conflicts and eliminate the pressure points and bottle-necks of unequal resource consumption.

5. *Opposing and eliminating pollution and resource depletion:* This principle is exemplified by the Kyoto Protocol and the reduction of CO_2 emissions.

6. *Complexity, not complication:* This tenet involves realistic and nature-enhancing solutions to environmental problems. It emphasizes systems thinking rather than fragmented approaches to environmental problem solving, as well as actions that are integrated within sustainable practices.

7. *Local autonomy and decentralization:* This principle, asserted largely before the globalization movement of the last quarter of the twentieth century, affirms value in the strengthening of local self-government and material and mental self-sufficiency. (Naess, (1973), pp. 3–6).

The third interrelated movement of increasing significance in the sustainability paradigm is Ecofeminism. If deep ecology is concerned with correcting the biases of the anthropocentric duality of humans and nature that lead to anti-ecological beliefs and practices, then ecofeminism focuses on androcentric man/woman relations. The argument stems from the age-old Aristotelian division that sees women as being connected with nature and exhibiting Dionyesian and emotional qualities, while associating men with the non-material, the rational, and the Apollonian. Thus ecofeminism seeks to identify, critically examine, and correct environmental injustices that result from the unthinking application of these gender-divisive views to the human world.

Any sustainability solution requires a progressive transformation of economy and society, the achievement of social equity between generations, provision for the basic needs of all, opportunities for people to satisfy their aspirations for a better life, and the achievement of sustainable consumption patterns. It requires economic growth in places where people's basic needs are not being met and the provision of equal opportunities. It also requires that demographic organization be put in harmony with the changing potential of the ecosystem, as well as the cessation of all forms of over-exploitation – including the over-exploitation of water reserves. According to Wilson (1998, p. 284), by 2025 over 40 per cent of the world's population will be living in regions of chronic natural water decline.

A litany of ecological policy objectives in response to increasing pressures from environmental depletion would seemingly rate highly on any liberal politician's public-good wish list, but achieving change in social organization, economic practices, and attitudes towards the environment is incredibly complex in practice. To begin with, there is frequently a communication gap between environmental lobbyists and emerging assistance groups, who may be unaware of their common interest and potential combined power. More frequently, there is simply a disconnection between political objectives and economic realities in nation-states, (particularly in the developing world), as well as a lack of education for change (WCED, 1987, p. 3).

One myth is that sustainability requires a reduction in consumption and a lessening of the quality of life. In fact, the first stage of sustainable practices requires growth, but growth of a different order – rather an enhancement of a different quality of life in order to make material- and energy-intensive practices more equitable in their impact. Advocates of sustainability maintain that rapid growth combined with deteriorating income distribution may be worse than slower growth combined with redistribution in favor of the poor (Brundtland, 1987, p. 3). The question then becomes, can sustainability be characterized in terms of social class and socio-economic inequality? The people who currently have the widest range of choices in lifestyle (for the most part, those in the developed world) also have the most freedom to make a difference by choosing sustainable lifestyles. Scratch the surface of environmental activism and many sustainability advocates believe that industrialization tends to simplify ecosystems (as well as degrade their quality) by reducing the diversity of species. New models of post-industrialization need to accentuate diversity and broaden the range of economic variables to satisfy human needs and aspirations by adjusting to scarcity through greater efficiency in the use of natural resources and by finding bio-sympathetic substitutions. However, the counterpoint to environmental degradation has been a stockpiling of biota, including genetic materials. In this context, Parry (as cited in Clark et al., 2006, p. 120) has pointed out that since 1980 "the world has witnessed the most significantly concerted ... accumulation of biological material since the nineteenth century."

As a movement of intellectual transnationalism, sustainability defies national boundaries, but there are insuperable tensions between the economies of profit and the economies of sustainability. Ironically, it is the developing world that is the greatest consumer of raw resources, while frequently demonstrating the fewest sustainable practices under the perpetuation of the neo-colonial paradigm. A sustainably literate first world can educate the developing world through sustainable economic intervention, through technology, and by development in the form of sustainable infrastructure. However, this requires an ideological involvement in sustainable practices extending beyond the repatriation of economic loans. It may do this by lobbying on the part of non-governmental organizations, by applying other forms of political and diplomatic pressure, and by its purchase choices in the market-place. At the grass-roots level of application, examples of practical measures for developed societies include waste sorting, eco-planting, solar panels and wind farms, alternative energy sources, and the diversion of a portion of industrial profits to the environment. Green mileage charges on air flights, the energy savings on eco-bulbs, and the use of hybrid cars are seen as the most likely small-scale solutions that aggregate into larger sustainability gains. These work in conjunction with technological invention and green management planning. Health care is also vastly dependent on ecology; 80 percent of the world's inhabitants still rely on traditional plant medicines for primary health care and 25 percent of prescription drugs contain extracts derived from plants (Clark et al., 2006, p. 122).

As well the use of new technologies to achieve social good, education may be used to define regimes of sustainable human development and can become a powerful tool for critiquing, disseminating, and archiving sustainability principles and practices. Objectives of sustainable development and environmental protection must be built into the mandates of the institutions that work in environmentally sensitive areas. The aim is to generate knowledge and expand the radius of influence of education for sustainability to create a base that fosters self-reliance. Concomitant with this is the need to establish a social system that provides solutions for the tensions that arise from disharmonious development. Such an administrative system needs to be flexible and have the capacity for self-correction. It may also require autopoietic qualities, i.e., be self-organizing with no inherent limits on its duration).

In exploring society and nature from a sociological perspective, Meadows, Meadows, Randers and Behrens (2005) ask whether there are physical and ecological limits to the expansion of the human population. One of the core issues of the Brundtland report (WCED, 1987) is to explore the notion of 'our common future' and in the context of sustainable development, to determine whether and to what extent the current generation is creating an environment that threatens the lives of future generations. However, in 1987 when the Brundtland report was published, it was thought that the assumptions behind the 'limits to growth' hypothesis would be ethically untenable. Strategists held the view that an ever-increasing population is not the real problem, but rather the emphasis that Western society places on materialism and consumption. In a nutshell, the problem can be reduced to the question of whether the main issue is the lack of resources in general, or whether the basic problem is that those scarce resources are held by the consumptive elites. These considerations involve two main factors: 1) effective citizen participation in decision making, and human rights, and the search for new solutions, including technological innovations, and 2) new organizing principles for distribution of (and access to) resources.

Edward Wilson (1998) defines two contradictory human attitudes in environmental debates. The first the exemptionist view, (which originates in the Christian interpretation of the Book of Genesis in the Bible), that Homo sapiens exists apart from nature and controls it, and humans can 'take or leave' the ecological view (Wilson, 1998, p. 278). The second is a naturalistic image in which humans are cradled within a razor-thin biosphere and are constituted in an organic 'habitat evolution.' As Flannery (2006) states:

> The atmosphere looks big because it is made of gas, but if we could compress that gas to liquid we would discover that the atmosphere is only one five-hundredth the size of the oceans. That's why humanity's major environmental problems – the ozone hole, acid rain, and climate change – result from pollution. And yet the atmosphere is dynamic. The air you just exhaled has already spread far and wide. The CO_2 from a breath last week may now be feeding a plant on a distant continent, or plankton in a frozen sea. (p. 18)

It has also been claimed that sustainable development is an oxymoron under a manageralist paradigm (Sveiby, 2009, p. 5); yet many management programs adopt a utilitarian approach to business innovation. Sustainability must be seen as more than part of an emerging paradigm of change and as driven by informed necessity rather than cosmetic or fashionable appeal. Yet where else may sustainable views as a set of attitudes that have underlying values be applied? How can these be adapted to business and education? Can society create sustainable economic and social systems, as well as sustainable teaching and management practices? Where can people turn to find some building blocks for models of sustainable practice?

Indigenous Models

Drawing on native Canadian Indian custom, Sveiby (2009, pp. 8–15) points to at least twelve aboriginal principles for sustainable development. These may be divided under categories of ecology, social factors, and economy:

Ecology:

- Keep all alive.
- Do not stay only in one place; (do not put all your eggs in one basket).
- Do not deplete the breeding stock.

Socialization:

- Do not impose your views on others.
- Share the knowledge.
- With knowledge comes responsibility.
- Divide the roles.
- Everyone has a role.
- Behave with responsibility to other communities.
- Punish only your own.
- If you break the law, you carry shame.
- Build respect.
- Maintain equitable power structures.
- Build community.
- Do no harm.

Economy:

- Knowledge is a primary resource.
- The economy has tangible and intangible parts – both are valuable in establishing quality of life.
- Respect diversity
- Do not sell products and services of low value

These general principles of environmental, social and economic behaviors and practices are coupled with concerns about the management of local resources, the stewardship of biodiversity, and the validity of local knowledge, all of which may complement scientific knowledge. However, it takes a shift in perspective for many schooled in the late twentieth-century and early twenty-first-century Western education system to trust the validity of knowledge acquired from sustainable use, rather than from empirical experiment and abstract deduction. In fact, two threats to some sustainable practices that require the expenditure of time and energy in terms of reorganization from a sustainability perspective are what are perceived as dogmatic and coercive education strategies concerned only with throughput efficiencies, and business practices that place profits over the welfare of the environment (applicable to problems of waste disposal, for example). Similarly, the holistic management styles favored by indigenous peoples in the sustainable paradigm mean more than recognizing the sum of parts of human interactivity with the environment. Rather, they require knowledge of the interrelationship between the parts. Holism is not a term associated with vague mythology, but rather a call to recognize the complexity in systems.

Traditionally, schooled aboriginal people see the environment as a complete system whose constituent parts are interconnected in a seamless web of causes and effects, actions and outcomes, behaviors and consequences, in which the past is linked with the present. People, animals, plants, and natural objects are not necessarily separate and distinct, but rather are linked and interrelated in a habitat. However, we must remember that indigenous peoples themselves are diverse the world over and their knowledge generation and sequestration is not entirely homogenous. It has been claimed that the notion of a 'public' is, after all, an eighteenth-century European concept, [see Immanuel Kant (version 1784), *What Is Enlightenment?*] Indigenous habitats are different in different regions. Nevertheless, the culture of the developed world has much to gain from collaboration with different indigenous traditions, methods, and techniques, not least the values and attitudes that represent local knowledge of the management of food production practices, traditional medicines, and the preservation of the genetic diversity of biota. Concepts about the usage of resources, attitudes towards human interrelation, communication, and the exchange of knowledge and information, as well as the techniques, strategies, and rationales for sustainable practices, may involve a confluence of Western techniques and indigenous societal lore in which there is a dialectical relationship between the practices and belief systems of different ethnic groups.

Four overriding sustainability principles from indigenous cultures are:

• respect for all life forms
• not to waste through play
• inter-generational knowledge transmittal, lifelong learning
• interconnectedness

According to Tipene (personal communication, January 28, 2009) Kaupapa Mäori concepts offer several principles for the guidance of sustainability in Aotearoa. These may include: a belief in *mauri* (a core essence or life force), the presence of *atua* (or custodians), the concept of *whanaungatanga* (belonging instead of owning), *whakapapa* (the lineage that connects Mäori to every aspect of the universe and each other), matauranga (knowledge and understanding – the concept of knowledge as identity, history, customs, genealogy and mythology), and *kaitiakitanga* (conservation and protection) (Jacobs et al., 2008). Tipene suggests that these generic concepts provide reference points for developing frameworks for self-sustainable land care and community resources. As part of the Integrated Mäori Land and Resource Development: A Decision-Support Framework ('Iwi Futures') project of Massey University, Tipene suggests that Kaupapa Mäori emphasizes the integration of a variety of cultural tools and practices, which may include te reo (language), knowledge, people, and land. These tools may be developed as methods of advancing economic, social, and ecosystemic integration and or fostering self-reliance. Principles and values underlying Mäori sustainability include: respect and caring (*kaitiakanga*), recognition of common interest (*whakapapa*), acceptance of responsibility (*tino rangitiratanga*), persuasion and empowerment (*manakitanga*), inclusiveness (*whanangatanga*), and equity (*oritetanga*).

Ecopsychology

Ecosystemic integration and self-reliance are also characteristics of the emerging discipline of ecopsychology, which is underpinned by a belief in the reciprocity of human–earth coexistence. The central tenet of ecopsychology assumes that it is psychologically damaging for humans to live in ways that are disconnected from their ecological context. At a person-centered level, ecopsychology is primarily concerned with methods of healing the disconnection of people from their ecological context through therapeutic techniques that involve such practices as mindfulness, daily ritual, heightened awareness, wilderness experience, and the development of a sense of place. These may be combined to produce ecological embeddedness – thus relieving symptoms of depression, stress, anxiety, longing, and grief. Ecopsychologists believe that mindfulness, the practice of being environmentally aware in the present situation (wherever that may be), contrasts with the fragmentary demands placed on people's attention in fast-paced industrial society and is one method of remaining ecologically connected. Interconnectedness at the individual level and a belief in the value of connectedness are relevant for sustainability because reconnecting humans to nonhuman nature is a step towards 'healing the planet'. Above all, ecopsychology attempts to remedy the anthropocentric stance that controls and objectifies nature regardless of human needs, exploring and neutralizing pathologies of destruction and disconnection.

At the other end of the spectrum of human and environmental utility/disconnection and conservation/connection, Edward Wilson (1984) defines the 'biophilia hypothesis.' Wilson's hypothesis holds that humans have an innate affinity for nature stemming from an ancestral past in which humans arose from the natural environment. Its antithesis is the 'biophobic' attitude born of ignorance, which separates humanity from the natural world and which ecopsychologists believe may be corrected by a nature-immersion experience. Ecopsychology is guided by a desire to encourage a human bond with bioregions, which provides a sense of belonging and motivates earth-friendly behavior.

Along with person-centered mindfulness, ecopsychologists advocate the wilderness experience (as proposed by Thoreau and others) as a means to ecological connection. This practice is augmented by adventure-challenge pursuits, which boost self-esteem, connection, and confidence through a sense of accomplishment at problem solving and overcoming physical challenges in natural environments. Interestingly, an orientation towards materialistic values (which focus on image, money, status, and possessions) tends to be negatively correlated with subjective well-being. This may be because consumerism and materialism are associated with individualism, which is at odds with collectivist values.

Ecopsychology also questions the relationship between phenomenology and empiricism. Exponents question the pursuit of logical-positivist inspired science, which has the corollary of estranging people from direct human experience. To regain a sense of mindful, ecologically connected physicality, ecopsychologists advocate perceptual awareness through a number of techniques, including: 'splatter vision' (which aims at widening the perceptual field), 'small world' (which focuses on enriching perceptions of environment), 'focused hearing' (the identification of sounds in an environment), the 'blindfold walk,' ' human camera,' and 'drum-sick blindfold walk' (which all increase perceptual capacity), and through 'scent trail tracking' (following a natural path according to smell) (Scott & Koger, 2005, pp. 5–6). Ecopsychology had its origins in the 1980s with the Council of Beings, a group designed by John Seed and Joanna Macy (early exponents in America of the deep-ecology movement) as a re-earthing workshop to help participants experience a connection to the natural world, both emotionally and spiritually.

Business and Sustainability

Sustainability is at its most contentious in the business environment. Despite the presence of strong regional, national, and international sustainability lobbying groups, (such as SANZ), the increasing concern of many Western governments to address issues of sustainability through national programs of education and legislation, and the growing presence of green agriculture, green energy, and a green consumer culture, it is very difficult to 'proof' the consumer-driven, profit-motivated economy with sustainability issues and practices. The methodological basis of business recalcitrance towards sustainability is logical positivism and concomitant philosophies of political skepticism such as those of Karl Popper. In his *The Open Society and Its Enemies* (1945), Popper claimed that open societies such as Western democracies, with their cornerstones of individual freedom, market economy, and competition, cannot implement solutions from what he terms 'closed' societies, such as primitive societies (which they surround). However, Popper's goal was to preserve democracy from dictatorship, not to assess the positive qualities of aboriginal societies. His use of the pejorative 'primitive' was largely reserved for European political abuses.

Nevertheless, exponents of sustainable practices argue that the traces of political ruthlessness are carried over into neo-liberal industrialization, in which a growing public awareness of a 'green economy' is countered by entrenched patterns of supply and demand for resources under the profit motive. In steering business towards a greener path, financial incentives and disincentives (including eco-taxes and tradable permits) that encourage certain kinds of economic behavior may be preferable (O'Sullivan & Painter, 2006, p. 14).

Sustainability advocates all point to the fact that the economic system is often perceived as independent of ecology, and interventions are often perceived as anti-growth – therefore anti-business. However, the economy, including the paradigm of creative industries and post-industrial stock markets (such as futures), is located in the confines of the ecosystem and cannot be separated from it. O'Sullivan and Painter (2006, p. 16) also consider the common belief that protecting the world's climate would *necessarily* have to be costly, as undemonstrated and questionable.

Elliot (2005, pp. 1–9) defines eleven factors that call into question the efficacy of sustainable business performance:

- Concern for sustainability issues is high, but the strength and depth of attitudes (level of belief) is low, which accounts for the perception that the rate of progress towards a culture of sustainability is less than desirable.

- Current methods of categorization of consumers that measure low levels of activity rather than underlying attitudes to behavior may under-estimate the level of commitment to sustainability,

- Questions that are biased in their formulation are inflating positive and negative responses.

- While levels of concern are high, in practice sustainability issues are not necessarily the highest priority for the general public, which means that potential activities that promote sustainable practices may be subordinated to other concerns.

- People in the corporate environment have no special access to information on sustainability.

- The level of debate about sustainability issues is uninformative, consisting of one-sided arguments, vested political interests, and exaggeration for effect, which all devalue the strength of the sustainability case.

- Information, though widely available, is not very accessible and many ingredients for cultural change are largely absent. If the level of belief is low, there is no effective feedback on achievement. There may be contradictory advice on appropriate actions and a lack of role models. It is hard to make the emotional connection that is needed for change to take place.

- People are unconvinced – they are fed up with an unfulfilled fear-mongering message.

- The green agenda is so vast and there are many issues over suitable terminology for describing it. Green labeling is confusing rather than helpful and may be applied indiscriminately.

- In purchasing decisions, product's benefits in terms sustainability are usually secondary to the main evaluation criteria.

- Marketers need to take a more holistic view. This doesn't mean seeing a totality, but not losing focus on the small things that matter. However, Elliot (2005, p. 6) claims that in most cases marketing based on a green or socially responsible platform remains ineffective. As he points out, companies tend to operate in one of three ways:

1. At low cost, with well-defined efficiencies, levels of service, deliverables, and price platforms.

2. With differentiation, defined by particular characteristics that add value for the customer, resulting in improved prices but pressure on margins as added costs.

3. With market focus, which means that a company targets a customer segment with selected characteristics to meet niche demand.

Often businesses have conflicts between providing minimum costs to consumers who buy on price and continuing to develop differentiating factors that may not be properly remunerated. The risk is that sustainability is seen as a differentiated consumable. This results in a so-called 'green-wash-syndrome,' attributed to packaging and advertising rather than applied throughout the business organization's management system as a guiding philosophy for best practice or for the composition of the end products (Elliot, 2005, p. 8).

There is also an element of relativity in the debate about the efficacy of green business. Surveys may be designed to produce different results, depending on business and political interests. For example, the US-based Natural Marketing Institute's 2007 LOHAS Consumer Trends Database reports a growing impact of consumer confidence in environmentally concerned businesses, but does not state whether its survey relies on a data pool of environmentally concerned consumers or is based on results from a set taken at random. (Survey results on the benefits of sustainability for purchaser loyalty and influence revealed mid-percentile increases in product purchase, consumer loyalty and good will, and lower increases in price concern).

Daly's (1996) rule for sustainable economic resource activity may be of more theoretical import than Elliot's assessment of market opinion is. As O'Sullivan and Painter (2006, p. 10) summarize, Daly's view is a form of 'input–output model' in which economics is directly related to the ecosystem. The *input rule* holds that:

1. Harvest rates of renewable resource inputs should fall within the regenerative capacity of the natural system that generates them.

2. Depletion rates of non-renewable resource inputs should be equal to the rate at which renewable substitutes are developed by human intervention. Any proceeds from sale of non-renewables should include research in pursuit of sustainable substitutes.

The *output rule* holds that waste emissions should be within the assimilative capacity of the local environment to absorb, without degradation to its capacity to continue to do so (Daly, cited in O'Sullivan & Painter, 2006, p. 10). The attractiveness of Daly's model is that it is eminently practical at the local level and also applicable within straightforwardly measurable parameters. However, others have also developed sustainable business policy statements.

Paul Hawken (1993, pp. 14–15) identifies eight principles and goals of sustainable engagement and development in business practices. In brief, these principles and goals are to:

- reduce absolute consumption of energy and natural resources in the industrial north by 80% within the next half-century
- take steps to provide secure, stable, and meaningful employment for people everywhere
- be self-actuating as opposed to regulated or morally mandated
- honor market principles – recognize that sustainable practices may not work if they require wholesale change to dynamics of the market. It is unlikely that people will respond to being asked to pay more to save the planet
- be open to the fact that sustainable practices may lead to societal rewards of a different qualitative form from the present way of life
- exceed primary sustainability objectives by restoring degraded habits and ecosystems to their fullest biological capacity
- rely on current income
- be fun and engaging, and strive for an aesthetic outcome.

Hawken further suggests six practical steps for ecological commerce, or how business can save the planet (1993, pp. 14–15). In summary, these are to:

- replace nationally and internationally produced items with products created locally and regionally
- take responsibility for the effects they have on the natural world
- not require exotic sources of capital in order to develop and grow
- engage in production processes that are human, worthy, dignified, and intrinsically satisfying

- create objects of durability and long-term utility whose ultimate use or disposition will not be harmful to future generations
- change consumers to *customers* through education.

These basic concerns may take their place in a wider debate about the benefits or disadvantages of globalization (the internationalization of business practices). They are straightforwardly positive tenets, although elaborate systems may need to be devised to apply them. However, in the first decade of the twenty-first century, there is also an increase in lobbying for international sustainability standards and collaboration among businesses over sustainability issues. Examples include a national industrial symbiosis program that has been launched in the UK and a growing movement towards corporate responsibility in America and the European Union following the global recession of 2008 (Waddock, 2008, p. 87).

The media profile of competing business concerns may be inherently tied to political concerns that perhaps form a company philosophy entailing a position on sustainability. Historically, people of liberal and democratic political leanings have favored green business while those of a conservative and republican persuasion have ignored it. Furthermore, under conditions of globalization, transnational production and movement of goods and services results in networks of supply and demand so varied and diverse that sustainable practices and products may make up only a small fraction of these. Underlying many national business conglomerates are government subsidies that support business expansion to maintain tax revenue. Employees of businesses in turn support expansion to keep jobs, but the consequences are income disparity and often environmental degradation.

If in the Western neo-liberal view, the global movement towards sustainability is informed by the systematic proliferation of sustainable management practices combined with the ecological knowledge of indigenous cultures, it may be unrealistic to place too much emphasis on local culture; yet the sustainability movement cannot overlook national, international, or local community interests. The tension between homogenization and diversity applies on all economies of scale. Sustainability has to be an inbuilt part of consumer choices, not just a vague ideological preference.

However, more and more business advice is offered from mandates for sustainability. Sustainability is also becoming synonymous in business planning with long-term success. The present UK government's business advice and support service, Business Link, identifies 17 factors for sustainability in its list of title guides (Business Link, 2009, pp. 1–4). The headings of these title guides may read like imperatives of the practical sustainability movement and confirm the relationship between the economy, society and the environment:

- Save Money by Reducing Waste
- Save Money by Using Energy More Efficiently
- Your Responsibilities for the Environment
- Importance of Environmental Issues to Your Business
- Set up an Environmental Management System
- How to Manage Waste Effectively
- How to Prevent Water Pollution
- Protect Employees and the Environment from Air Pollution
- How to Use Environmental Assessment Techniques
- How to Make Your Supply Chain Greener
- Your Responsibilities for Health and Safety
- Use Resources More Efficiently
- Grow Your Business Through Sustainable Innovation
- Provide Sustainable Goods and Services
- Corporate Social Responsibility
- Ethical Trading
- Create a Strategic Approach to Sustainable Development

In these guides, sustainability applies to values, attitudes, processes, practices, people, goods and services, communication, and transportation. However, sustainable business practices may also have deeper ecological connections in the context of appropriating designs from nature and natural systems. Quite separate from the fact that human emotions may correspond to physical states in nature, the concept of biomimicry or biotechnics points to ways in which sustainable technology can be designed and engineered from the implicate order of nature.

As Benyus (1997) points out, biomimicry uses nature as a model. It is innovation based on patterns discovered in nature and adapted into designs and processes that solve human problems. Examples are the solar cells (inspired by a leaf design), suspension bridges (inspired by skeletal structures), velcro (inspired by grappling hooks on seeds), and even computer inter-connectivity (mirrored by hymenop-terac society). The concept also involves nature as a measure of human endeavor, with the tacit knowledge that natural designs have evolved over 3.8 billion years and are, to some extent, reasoning from induction – therefore, self-sustainable in evolutionary terms, (despite the second law of thermodynamics, which the physicist Schrödinger had identified in 1943 as being quite separate from life forces).

Biomimicry may be applied to resource management systems as well as to technological inventions. An example is the cleaning of water systems by passing the water through ponds of gravel and reeds, as is done at Wellington City's Waitangi Park in New Zealand. The final cornerstone of biomimicry is the need to study and learn from natural designs and systems in new ways, asking not only what can we extract from nature, but what can we learn and apply from it.

Market-Based Solutions

Green consumerism holds that market processes and the consumption decisions of millions of individuals can themselves create an environmentally sustainable society. A central idea of the market economy is that as one kind of resource starts becoming scarce, its price will go up. Yet many sustainable resources, some of which we might take for granted (such as water, air and food), are needed by all. As we already know, the quality of life is uneven over much of the world. Western society can play a leadership role in sustainable business practices and at the same time teach and provide innovative technologies for sustainable systems in the developing world. The alternative will be a doomsday scenario of societies that are politically organized for the predation of basic human resources.

A counterpoint to this is the unproven belief that the open market may be a self-balancing system. If the market is analogous to an ecological system, can solutions be found by relying on local knowledge and the intelligence of individuals rather than on state intervention? One libertarian theory expounds a form of 'hands off' intervention along the lines of Adam Smith's notion of the market delivering a regulated economy by means of the 'invisible hand' of supply and demand. Yet it is easy to see this as an excuse for the individual to do nothing. How viable is the concept of effortless ecological modernization? Science and technology may have been responsible for getting modern societies into a poor environmental state, but with education for sustainable practices, they are equally capable of getting societies out of that state and into a new environmentally stable modernity. Under a regime of ecological modernization, environmental standards may gradually be raised (Beck, 1982, p. 3). An example of 'ecological modernization' is the business movement towards green products. These include organic products whose manufacture is achieved through environmentally friendly processes. Such products may or may not be identified on store shelves by 'green labeling'. For example, a shirt for sale at Fair Trade in Wellington, New Zealand, made by a tailor in Kathmandu, Nepal, may be purchased over the Internet from almost anywhere in the world.

Beck also suggests that we live in an age of manufactured risk: risk is no longer an act of God, but of science-based intervention in the natural world (solving social and economic problems). As such, science and politics are our 'second modernity' (Beck, 1982, p. 4). But what of morality? Beck further suggests that new forms of reflexivity are developing in which people are losing faith in all forms of authority (including that of scientific enterprise), and are creating their own understandings. Increasingly, at least in the developed world, the 'new individualism' is contested by the notion of a global citizenship. This includes the proliferation of green technologies on a local scale. However, this is counterbalanced by the need to maintain communities of sustainable integration. In Brown's article "It's income tax time for Americans, and it's time for the entire world to lower income taxes and raise environmental taxes," he warns that socialism collapsed because it did not allow the market to tell economic truth, and predicts that capitalism may collapse because it does not allow the market to tell the ecological truth (Brown, as cited in Peet, n.d., p. 16).

Education

'Education for sustainability' is a term developed from the natural and social sciences. It denotes the process of change that leads people and communities to live in sustainable ways. In its role as critic, conscience and arbiter of social good, higher education has a crucial role in analyzing, synthesizing, authenticating and disseminating the message of sustainability. The main goals for sustainability in education include:

- bringing about changes in behavior and lifestyles
- disseminating knowledge and developing skills
- incorporating sustainability in pedagogy
- incorporating sustainability in management practices and processes
- preventing the exhaustion of non-renewable resources.
- equity and fairness
- education for well-being and peace education
- rights of indigenous cultures
- integration

Stone and Baldoni (2006, p. 1) claim that the sets of expectations for sustainability in education are multiple and include: revising the underlying values of society, revising the nature of programs themselves incorporating sustainability templates, widening the pool of knowledge about sustainability, targeting specific audiences for sustainability education, mandating institutional requirements for sustainability, and broadening and deepening educational research and development. Sustainability in pedagogical practices includes the analysis of the interactivity between social, economic, and environmental concepts, as well as knowledge sharing and environmental awareness. Relevant curriculum areas are: environmental law, diversity in cultural and ethnic perspectives, ecological life-support systems, methodologies in research and investigation, the interdependence of systems, organizations and relationships, the maintenance of biodiversity, environmental literacy, climate change, and the cultural and population demographics within the global geographica. Methods of curriculum engagement in the sustainability movement include:

- partnerships, cooperation, collaboration
- integrative and cross-sector learning
- futures thinking, situation improvement, and making judgments about quality gains
- critical enquiry, reflective thinking, contrasting, generating solutions
- cooperative learning, enquiry-based learning, experiential learning.

Stone and Baldoni (2006) also acknowledge that sustainability is most clearly recognized in engineering technology programs. They further conclude that concerns in tertiary education vary from understanding natural and physical processes to knowledge dissemination and pedagogy in multidisciplinary and resource management curricula.

Calder and Clugston (2003, p. 4) advocate the encouragement of all tertiary educational organizations to engage in education, research, policy formation, and information exchange on sustainability issues and practices. Furthermore, they propose the establishment of programs to produce expertise in environmental management, sustainable economic development, and population demographics, in order to ensure that all university graduates are environmentally literate and have the knowledge to become ecologically responsible citizens. Certainly this is consistent with the political rhetoric coming from the 2009 US Democratic administration of Barack Obama, with its pre-election promise of a $150 billion Apollo Alliance Project to bring green jobs and energy security to the USA through an alternative energy economy (Lean & Doyle, 2008). These green jobs will require training for both industry and the education sector.

Calder and Clugston (2003) further emphasize the importance of new pedagogical approaches that include systems thinking, learner exposure to issues of equity and justice, and interdisciplinary learning. They advocate the establishment of curriculum courses involving sustainability topics such as: globalization and sustainable development, urban ecology and social justice, women's development, campus ecosystems (sources of food, water, energy, waste endpoint), renewable energy, sustainable building design, ecological economics, populations and development, environmental justice, reducing the ecological footprint, sustainable building construction and renovation, environmentally responsible purchasing of food, uses of consumables (paper other products), student orientation, lecturing according to sustainability issues, and fostering local, regional, and global partnerships. They further call for TEOs (Tertiary Educational Organizations) to include sustainability in organizational charters and mission statements, incorporate structural procedures into mission plans, and register and communicate sustainability through environmental reporting mechanisms (Calder & Clugston, 2003, p. 5).

This orientation towards sustainability is consistent with the *Talloires Declaration 10 Point Action Plan*, an agreement signed in 1994 by an Association of University Leaders for a Sustainable Future that

calls for urgent actions to address problems of "unsustainable production and consumption patterns that aggravate poverty in many
regions of the world" (Association of University Presidents for a
Sustainable Future, 1994, p. 1). The goals of the Talloires plan are to
educate, research, form policy, and exchange information in order to
achieve "stabilization of human population, adoption of environmentally sound industrial and agricultural technologies, reforestation,
and ecological restoration" (Association of University Presidents for
a Sustainable Future, 1994, p. 1).

Calder and Clugston's concepts and the Talloires Declaration all
point to the critical role in higher education of authenticating, legiti-
mating, and disseminating the sustainability message as organizations
that reflect the priorities of the societies in which they take part. Institutions adapt to the demands of government and disciplines are
shaped by what academics, funders, and stakeholders require. Sustainability involves shifting the scope of these requirements towards
a set of achievable goals that are socially, economically and ecologically informed.

In the education wing of The Tahoe Center for a Sustainable
Future (California), Goldman et al. (1999, p. 11) has identified 17
main points for teaching sustainability:

- strong core academics
- understanding relationships between disciplines
- systems thinking
- lifetime learning
- hands-on experiential learning
- community-based learning
- effective use of technology
- partnerships
- family involvement
- personal responsibility
- human development and earth's natural system
- increased awareness of local, environmental, cultural, and economic issues
- stewardship of environment, personal connection to social and
  environmental aspects of their community
- increasing comprehension through work with positive adult role
  models

- develop critical thinking skills
- develop citizenship skills – diverse view points
- systems thinking: connections, concepts, relationships to whole, material and intellectual processes, interactive consequences.

Applying Turner, Ignace and Ignace's (2000) view of traditional ecological knowledge to the education system is a matter of correlating similarities within processes and resources. The succession and interrelatedness of all components of the environment corresponds to courses/degree structures, and interdisciplinarity. The use of ecological indicators and adaptive strategies for monitoring, enhancing and implementing sustainability education corresponds to managerial processes. Effective systems of knowledge acquisition and transfer, and respectful, interactive attitudes and philosophies all correspond to research practices. Close identification with ancestral lands and beliefs that recognize the power and spirituality of nature correspond to the ecology of the TEO base.

As Dale and Newman (2005) have suggested, a common criticism of sustainability in higher education is that it is normative, ambiguous, and ineffective. This may be due the fact that it is values-based as well as empirically motivated in bringing about sustainable practices. However, sustainability is rarely criticized for being unreflective; it implies a degree of analytic competency and proactive systemic thinking. Inherent in all sustainable motivations is a belief in inter-generationalism: solving problems for the current and future generations. Most programs for educating sustainability will claim that pedagogical aims include acquiring a set of skills to reflect upon the complex nature of social, environmental, economic, and systemic problems that involve adaptive systems thinking. Skill sets include: interdisciplinary interconnectivity, environmental literacy, epistemological enquiry, facts-based analysis involving individual situations and group relations, and the ability to fit the local and specific into a global whole.

As Dale and Newman point out, sustainability is one of the few disciplines in which vagueness has a purpose – the ability to conceptually adapt and modify to new situations, remaining flexible to change and accommodating to resistant, shifting value systems (Dale and Newman, 2005, p. 2). Waldrop (1992) maintains that complex systems have two defining properties: self-organization, and adaptation to change. Bar-Yam (1992) theorizes that human civilization is a super-organism. While natural capitalists may be attuned to the physics of the economy Hawken (1993), suggests that sustainability practitioners as 'natural capitalists' explore the reconciliation of the value-chain of supply and demand, resource usages, and efficiencies that preserve naturally occurring value and biotic diversity under a rubric of interconnected ecological and social imperatives. As such, sustainability pedagogy may be informed by a process-orientated practice that represents a constantly changing set of objectives related to environmental usage rather than a finite goal (Dale & Newman, 2005, p. 3). While accommodating and promoting diversity, sustainability advocates also promote environmental unification.

As Wals and Jickling (2002) suggest, education for sustainability engages students in social-scientific pursuits. While the discipline may contain an inherent flaw in the concept of continuous sustainment (given the second law of thermodynamics as it applies to classical physics), the perpetuation of the human species within a sustainable biosphere is a fundamental human value as well as an evolutionary and physiological imperative.

Ecolinguistics

Language is inherently related to environment. Environment is both outside and within people. As Sapir (2001) suggests, the social forces that affect people may transform environmental experiences (pp. 13-14). Despite the fact that environment is often seen as exterior to human influence, there is a firm case that it can actually be seen as a porous boundary of which people are a part: "Yet speaking of language, which may be considered a complex of symbols reflecting the whole physical and social background . . . . it is advantageous to comprise within the term environment both physical and social factors" (Sapir, 2001, p. 14). Language is materially influenced by the environmental background of speakers but language also shapes the way people view the environment. It could be argued that people create more things in their minds than there are things in nature, but the adaptability and flexibility of language is a contentious issue. As Sapir (2007) states, "A large number, if not most, of the elements that make up the physical environment are found universally distributed in time and place, so that there are natural limits set to the variability of lexical materials in so far as they give expression to concepts derived from the physical world" (p. 17). However, people interact with the environment and thus create meanings influenced by their experience of it, sharing meanings with "physiological factors of a subtler character . . . operat[ing] in a diffusion of cultural elements" (Sapir, 2001, p. 19). As Chawla (2001) states, "The 'real world' is to a large extent unconsciously built upon the language habits of the group" (p. 116). More importantly for understanding eco-linguistics is the way in which language may be inherently diffused with concepts and terms that implicitly conceal ideologies, politics, and power structures that preserve the human and nature divide.

As Muhlhauser (2001, p. 31) suggests, three more obvious ways in which language may preserve meanings which are detrimental to ecological interests are through: a) semantic vagueness, for example: pollution, primitive, safe, deterrent, and b) semantic undifferentiation: a single term covering a spectrum of qualitatively different phenomena, and c) misleading encoding, for example zero growth. As Fill and Mulhausler (2001) states, ecosystems and languages are connected in so much as the former are life-systems and languages are systems of experience. Consequently, eco-linguistics is a critical study of language practices in terms of the relationship between people and environment. Ecosystemic thinking is thus more akin to cognitive processes and interpersonal communication than it is to the utilitarian machine metaphor.

Gerbig (as cited in Fill & Mulhausler, 2001, p. 46) points to two factors of language use found in expressions of environmental counter-interest groups. These are the collocational frequency of phrases concerning the "lexemes cause and responsible," and the suppression of agency through choice of "active, passive and ergative constructions" (ergative verbs are those which suggest a change of state). For example, ergative expression such as 'cases of water-poisoning have been increasing' hide the agency and imply a self-cause, implying that the action is the cause rather than human agency. Three other forms of language use may hide agency. These are: a) agency which implicates everyone 'our pollution', b) the deletion of agency through passivisation 'meeting minimum levels,' or c) nominalization – placing reduced emphasis on the affected 'land annexation.' Consequently, both agency and deixis (word utterances requiring contextual information to understand meaning) may be selectively suppressed in discourse concerning ecological issues. Furthermore, the "unlimitedness of our resources and the special position of humans" may also be interwoven throughout the language system (Fill & Mulhausler, 2001, p. 48). This is evident in the pronominal system: he or she is used for humans and the neutral 'it' for non-human beings.

Similarly, animals and plants are referred to through the occlusion of collocations involving experience (think, know, empathy, etc.) (Fill & Mulhausler, 2001, p. 48).

Fill and Mulhausler (2001, p. 48) also points to three other ways in which ecological interests may be compromised by language: firstly, by the frequent use of uncountable or mass nouns with regard to the infinity of natural resources (oil, air, energy, water etc.) , and secondly, by the use of contrasts in which the 'growth word' is always the neutral term (how high is the building not how low), and thirdly, by the way in which western language is non-admitting of agency from non-humans (thought is not ascribed to animals in any meaningful sense). Chawla (2001) points out that other cultures such as Native Americans exhibit a tendency to individualize mass nouns, distinguish between real and imaginary nouns, and use verb forms which treat time as fluid rather than having a three-dimensional form, whereas the English language requires the speaker to refer to "a physical thing as a binomial that splits the reference into a form-less item and a form" (p. 116).

The larger point to be seen here is that Standard Average European languages (SAEs) tend to be rooted in the Newtonian paradigm and tend to be unmodified by an ecological understanding. This fact questions the division of human agency in terms of spatial-temporal experiential positioning, emphasizing instead a divisional and fragmented view of the universe. Furthermore, SAEs frequently defy holism or an undivided wholeness instead dividing language meaning into "agentive participants, affected participants, and circumstances" (Fill, 2001, p. 49). The question remains then, should linguists actively partake in language re-adjustment towards ecological correctness, or will language use change in this manner over time? Such a process indicated by the use of ergative constructions and terms which express mutuality would accord with a deep ecologization of language (Fill & Mulhausler, 2001, p. 50). As Chawla (2001) suggests, "The modern technological world view does not adequately take into account the idea that all life on earth is fundamentally the same, that most differences which seem important to us are superficial" (p. 117). Ideally, language use and linguistic diversity should reflect biological diversity, but also the favorable usage of ecological resources. Inherent in this is the understanding that human behavior involves both outward action in the environment and also inward competence – understanding of a relationship with ecology. However, as Schultz (2001) suggests, in Western society, the language of common use often reflects that of commercial users of the environment. The language use agenda is frequently set by government, bureaucracy, industry, and the news media. However, many conservationists may be unaware of the way in which language use might carry implicit messages of environmental exploitation. One example is the use of 'weed' to describe trees that are deemed exploitable or 'waste' to mean those not suitable for exploitation; such terms negate the intrinsic value of such species to the ecosystem and are denigrating of any 'wild' species or those without foreseen commercial use. Schultz (2001, p. 110) points out that even the term 'sustainable development' may imply 'sustained development,' which is clearly unsustainable; rather, the meaning suggested is 'ecologically sustainable development.' Such linguistic constructions may also deny agency to ecologically defined issues, promote the diffusion of responsibility, and obscure the fact that most environmental problems do not follow from individual use but from exploitation by human institutions (Penman, 2001, p. 228). Finke (2001, p. 87) suggests that a language which has evolved from an ecology inherently has an openness, a hidden equilibrium, and pluralistic qualities, but then comments that what 'we' have made it is characterized by ambivalence and implicit values of politics (power), administration (order), or economics (business values).

Figure 1. Sustainability – The convergence of praxis

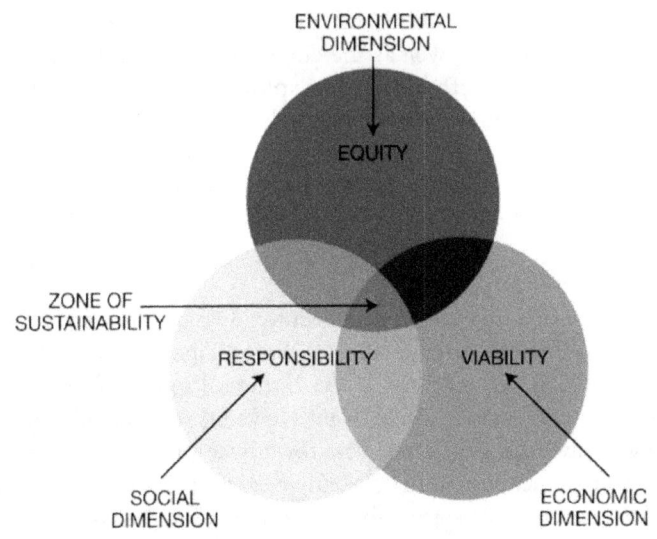

Conclusion

The sustainability movement calls upon the shared connections across and between disciplines to provide different techniques for registering and ameliorating the different views between the incompatible value systems that proliferate in the social, economic, and ecological spheres, leading to new formulations of sustainable practices at their confluence (Dale & Newman, 2005, p. 2). The inherent pedagogical aims of students of sustainability are to explore diversity and understand differences, to acquire the tools of analytical reasoning about the environment, the economy and society to mediate between contesting value systems, and even to hold two or more contradictory positions at once – culminating in the knowledge of how to act as informed agents for change towards sustainable lifestyles. While it is questionable whether the mantra connotes unconditionally positive values (Dale & Newman, 2005, p. 2), positive values have a necessary place in maintaining well-being and advancing motivational conditions for a sustainable human environment. Others have claimed that environmental issues are not fundamentally about sustainability, but about cultural identities, social and ecological equity, and relationships between society and nature (Dale & Newman, 2005, p. 2). The constant variable common to both these approaches in the disciplinary development of sustainability is that, given that human society and the biosphere are dynamic, it a process in which nothing is predetermined. Self-environment actualization may be a continual process of adaptation and change in response to the human and natural cycles, as well as the environmental, social, and economic pressures that determine it. An exploration of sustainability as it is related to indigenous principles, beliefs and culture, business tenets and philosophies, and education demonstrates how interdisciplinary inter-relatedness may result in an ethos of human environmental enhancement.

Acknowledgements
I am grateful to Open Polytechnic Executive Director Erima Henare for guidance on Māori sustainability, to Percy Tipene of the Massey University and Landcare Research Iwi Futures Project for guidance on Māori Kaupapa issues, to Dr. Kate Hewson, sustainability advocate at the University of Canterbury, for advice on sustainability issues in tertiary education and Dr Jonathan Barrett of the Open Polytechnic for discussion of sustainability and human rights issues.

References

Archer, D. (2010). *The long thaw*. Princeton, USA: Princeton University Press.

Association of University Presidents for a Sustainable Future. (1994). *The Talloires declaration: 10 point action plan*. Retrieved from http://www.ulsf.org/pdf/TD.pdf

Bar-Yam, Y. (1992). *Dynamics of complex systems*. Reading, MA: Addison-Wesley.

Barrett, J. (2010). *Human rights, sustainability and standing: A humanist perspective*. Lower Hutt, New Zealand: The Open Polytechnic.

Beck, U. (1982). *The risk society*. London, England: Sage Publications.

Beckerman, W. (1994). 'Sustainable development: Is it a useful concept?' *Environmental Values, 3*(3), 191–209.

Benyus, J. M. (1997). *Biomimicry: Innovation inspired by nature*. New York, NY: William Morrow.

Bourdieu, P. (1986) The forms of capital. In J. Richardson (Ed.) *Handbook of Theory and Research for the Sociology of Education* (pp. 241–258). New York, NY: Greenwood.

Business Link. (2009). *Make your business more sustainable: An overview*. Retrieved from http://www.businesslink.gov.uk/bdotg/action/layer?topicId=1079404746

Calder, W., & Clugston, R. (2003). 'Progress toward sustainability in higher education'. *Environmental Law Reporter, 33*, 10003–10023. Retrieved from http://www.ulsf.org/pdf/dernbach_chapter_short.pdf

Chawla, S. (2001). Linguistic and philosophical roots of our environment crisis. In A. Fill and P. Muhlhausler. *The Ecolinguistics Reader: Language, Ecology and Environment* (pp.115-223) London: Continuum.

Clark, N., Massey, D., & Sarre, P. (2006). *A world in the making*. Oxford, England: The Open University.

Dale, A., & Newman, L. (2005). 'Sustainable development, education and literacy'. *International Journal of Sustainability in Higher Education, 6*(4), 351– 362.

Daly, H. (1996). *Beyond growth: The economics of sustainable development*. Boston, MA: Beacon Press.

Elliott, D. (2004). 'Sustainable motivation: Attitudinal and behavioral drivers for action'. In *Responsible marketing 2004: Sustainability, quality of life and commercial success* [Collected conference papers]. Amsterdam. The Netherlands: ESOMAR.

Elliott, D. (2005, April). *The role of marketing at the business/consumer interface.* Paper presented to the 4th International Conference on Business and Sustainable Performance, Aarlborg, Denmark. Retrieved from http://www.mpgintl.com/sustain/english/report.htm

Fill, A. (2001). Ecolinguistics: State of the Art 1998. In A. Fill and P. Muhlhausler. *The Ecolinguistics Reader: Language, Ecology and Environment.* (pp. 43-53). London: Continuum.

Fill, A., & Muhlhauser, P. (2001). *The Ecolinguistics Reader: Language, Ecology and Environment.* London: Continuum.

Finke, P. (2001). 'Identity and Manifoldness: New Perspectives on Science, Language, and Politics'. In *The Ecolinguistics Reader. Language, Ecology and Environment,* ed. Alwin Fill and Peter Mühlhäusler. London: Continuum.

Flannery, T. (2006). *We are the weather makers: The story of global warming.* Melbourne, Australia: Text Publishing.

Goldman, H., Irelan, S., Fornais, C., Ritter, S., Rockwell, T., & Segale, H. (1999). *Sustainability education: Teaching sustainability in every classroom. An introduction and guide for educators.* The Tahoe Centre for a Sustainable Future. Retrieved from: http://ceres.ca.gov/tcsf/seg/ index.html

Gore, A. (2006). *An inconvenient truth* [Motion picture]. United States: Paramount Pictures.

Greenspan, A. (2007). *The age of turbulence: Adventures in a new world.* New York, NY: Penguin.

Hawken, P. (1993). *The ecology of commerce: How business can save the planet.* London, England: Weidenfield & Nicolson.

Hamilton, C. (2010). *Requiem for a species. Why we resist the truth about climate change.* Crows Nest NSW, Australia: Allen and Unwin.

Jacobs, D., Price, A., Amir, I., Maresca, G., Watterson, W., Edmeades, M., et al. (2008). *The outlook for someday: A Māori perspective on sustainability.* Retrieved from http://www.theoutlookforsomeday.net/assets/pdfs/The%20Outlook%20for%20Someday%20-%20A%20Maori% 20Perspective%20on%20Sustainability.pdf

Kant, I. (1784). *What is enlightenment?* Available from http://www.english.upenn.edu/~mgamer/Etexts/kant.html

Karl-Henrik, R. (2002). *The natural step theory.* Gabriola Island, Canada: New Society.

Lean, G., & Doyle, L. (2008). 'Obama's green jobs revolution'. *The Independent.* Retrieved from: http://www.independent.co.uk/

news/world/americas/obamas-green-jobs-revolution-984631.
html

Lomborg, B. (2001). *The skeptical environmentalist.* Cambridge, England: Cambridge University Press.

Lovelock, J. (2000). *Gaia, a new look at life on earth.* Oxford, England: Oxford University Press.

Meadows, D. H., Meadows, D. L., Randers, J., & Behrens, W. III. (2005). *Limits to growth. The 30-year update.* London, England: Earthscan.

Ministry for the Environment. (2003). *Sustainable development for New Zealand programme of action.* Retrieved from http://www.mfe.govt. nz/publications/sus-dev/sus-dev-programme-of-action-jano3. html

Ministry for the Environment. (2009). *Sustainable water programme of action (2003–2008).* Retrieved from http://www.mfe.govt.nz/ issues/water/prog-action/

Moffatt, I., Hanley, N., & Wilson, M. (2001). *Measuring and modelling sustainable development.* Lancashire, England: Parthenon.

Naess, A. (1995). 'The shallow and the deep, long-range ecology movement: A summary'. In A. Drengson & Y. Inoue (Eds.), *The deep ecology movement, an introductory anthology* (pp. 3–6). Berkeley, CA: North Atlantic.

O'Sullivan, A., & Painter, D. (2006). *Advancing sustainability through university academic formation – Experience with a professional engineering programme.* Wellington, New Zealand: Parliamentary Commissioner for the Environment.

Peet, J. (n.d.). *Sustainability in New Zealand – Lifting the game* [Sustainability review background paper]. Retrieved February 8, 2009, from http://www.pce.govt.nz/__data/assets/pdf_file/0014/ 1634/sustain.pdf

Penman, R. (2001). Environmental Matters and Communication Challenges. In A. Fill and P. Muhlhausler. *The Ecolinguistics Reader: Language, Ecology and Environment* (pp. 143-153). London: Continuum.

Popper, K. (1945). *The open society and its enemies: Volume 2: Hegel and Marx.* London, England: Routledge & Kegan Paul.

Sapir, E. (2001). Language and Environment. In A. Fill and P. Muhlhausler. *The Ecolinguistics Reader: Language, Ecology and Environment* (pp. 13-23). London: Continuum.

Scott, B. & Koger, S. (2005). Teaching psychology for sustainability: A manual of resources. *Ecopsychology*. Retrieved from http: http://www.teachgreenpsych.com/

Stern, N. (2006).*The economics of climate change*. Cambridge, England: Cambridge University Press.

Stone, L. J., & Baldoni, M.-J. (2006). *Progress and pitfalls in the provision of tertiary education for sustainable development in New Zealand. Sustainability Review: Background Papers*. Wellington, New Zealand: Parliamentary Commissioner for the Environment. Retrieved from http://www.pce.parliament.nz/assets/Uploads/Reports/pdf/te rtiary.pdf

Sustainable Aotearoa New Zealand Inc. (2009). *Strong sustainability for New Zealand: Principles and scenarios*. Auckland, New Zealand: Author. Retrieved from http://www.phase2.org/documents/ Strong_Sustainability_for_New_Zealand_v1_May_2009.pdf

Sveiby, K. (2009). Aboriginal principles for sustainable development. As told in traditional law stories. *Sustainable Development, 17*, 1–27.

The benefits of sustainability: Purchase, loyalty and influence. (2008, January 22). *Environmental Leader*. Retrieved from http://www .environmentalleader.com/2008/01/22/the-benefits-of-sustainability-purchase-loyalty-and-influence/

Turner, N. J., Ignace, M. J., & Ignace, R. (2000). 'Traditional ecological knowledge and wisdom of aboriginal peoples in British Columbia'. *Ecological Applications, 10*(5), 1275–1278.

Waddock, S. (2008). 'Building a new institutional infrastructure for corporate responsibility'. *The Academy of Management Perspectives, 22*(3), 87–108.

Waldrop, M. (1992). *Complexity: The emerging science at the edge of order and chaos*. Toronto, Canada: Simon & Schuster.

Wals, A., & Jickling, B. (2002). 'Sustainability in higher education'. *International Journal of Sustainability in Higher Education, 3*(3), 221–232.

Watson, P. (2000). *A Terrible Beauty*. London, England: Weidenfeld & Nicolson.

Wilson, E. (1984). *Biophilia*. Cambridge, MA: Harvard University Press.

Wilson, E. (1998). *Consilience: The unity of knowledge*. New York, NY: Alfred A. Knopf.

World Commission on Environment and Development. (1987). *Our common future*. Oxford, England: Oxford University Press.

3

ECOLOGICAL COMMUNICATION
The Anthropomorphic Bias

Introduction

This chapter provides a synthesis of contemporary research about anthropomorphic bias. It provides an explanatory distinction between the concept of anthropocentrism and the more specific applications of anthropomorphism, as well as a critique of existing anthropocentric concepts and theories. The chapter enters into analysis of some of the philosophical assumptions behind the related concepts of anthropocentrism and anthropomorphism. It provides a critical survey of the current literature and makes the argument that anthropomorphic bias can be understood as an innate existential tendency of human embodied thought, thereby presenting a potential problem to the fields of the philosophy of science and embodied cognition, and to social scientific experimental design and interpretation.

The chapter is divided into eight sections dealing with selected areas of discussion of anthropomorphic bias, involving summary explanations of experimental situations and everyday life behaviors: (1) Anthropocentrism and anthropomorphism: Definitions, ontologies, problematics and reflections; (2) Anthropocentrism and ecology; (3) Deep anthropocentrism and counter-environmentalism; (4) Anthropomorphism and human and animal differences; (5) Anthropomorphism and quantum physics; (6) Anthropomorphism and robotics; (7) The problem of anthropomorphism; and (8) Attitudinal solutions to anthropocentric bias involving new attitudes in scientific and everyday life behaviors. These areas of focus are chosen as they comprise the clearest categories of research for the concept of anthropocentrism at the current time.

What are Anthropocentrism and Anthropomorphism?

Anthropocentrism is a term used to describe the condition of being humankind-centric, emphasizing the need to protect the interests of humans and thus assessing the importance of other ecological entities in terms of their usefulness to humankind. Although the concept of humanity is constantly evolving, the term may carry a normative value. Anthropomorphism is a descriptive term used largely to ascribe human properties to non-human things, such as cars, ships, buildings, and landscapes. Anthropormorphism denotes specific instances and usages of the anthropocentric concept. In this chapter, the terms will not be used interchangeably, but rather denote either broader conceptual meanings or more specific applications. Anthropomorphism is also sometimes used in the context of human-like descriptions of animals, which theoreticians of embodied cognition regard as existing on the same continuum with human beings (Anderson, 2003). However, the ascription of animal-like qualities to animals is known as zoomorphism.

Although cognitive linguists such as Lakoff and Johnson (1980, p. 4) have argued that reason is evolutionary and should not be considered as an essential characteristic that separates us from other animals, the anthropic view can be conceived of as one of the emergent properties of being that are endemic to humanity. Arising from anthropological and environmental discourse, anthropomorphism has informed human culture and thinking from the moment of collective and individual awareness of humankind as a distinct species. Denis Diderot referred to it in *The Encyclopedie* (Diderot, 1755/1992). The contention is that community, society, and consideration of ourselves as humans necessarily predisposes us to an anthropocentric bias, which, when considered from different perspectives, may be both beneficial and harmful to ourselves and to society.

Anthropomorphic bias is a second-order problem in the field of embodied cognition. If the focus of research into embodied cognition (Anderson, 2003) emphasizes that cognition is a highly embodied or situated activity and that thinking beings are acting beings, then anthropomorphic bias arises as a consequence of our consciousness of selves and others as embodied beings. It emerges as an abstraction of our embodied cognitive states. For example, anthropocentrism may underlie our thinking at a level beneath that of motivated reasoning. Motivated reasoning occurs as a mental bias that follows from human reasoning. Ziva Kunda (1990) points out that:

> People are more likely to arrive at those conclusions that they want to arrive at . . . unrealistically positive views of oneself and the world are often adaptive . . . [b]ut motivated illusions can be dangerous when they are used to guide behavior and decisions, especially in those cases in which objective reasoning could facilitate more adaptive behavior. (p. 495)

Kunda's quote establishes a counterpoint to the notion of 'objective reasoning' by pointing out that motivated illusions may be dangerous if they are used to guide behaviors inspired by irrational self-beliefs. Leaving aside the question of motivated objective reasoning, anthropocentrism still points to a deeper problem. Anthropocentric predispositions exist at a deeper level than motivated illusions that support people's foregone, desired conclusions; it also denotes an influence on thought and behavior that is not necessarily what people want to do, but is just what they *do* do (or can't not do).

Anthropomorphic bias is a potential problem for both the classical Newtonian view of the world and for quantum mechanics because its influence on human thought and social organization is largely scientifically unaccounted for, yet which is nevertheless causally, (but not necessarily logically), unavoidable. More research in identifying types of anthropocentric behavior may make us more aware of how it affects us as individuals and as cultures, and place rational presuppositions on what we can and can't avoid doing while acting and thinking in human ways. It may also reveal hidden influences on people's objective reasoning that may be both beneficial and harmful in different circumstances.

The argument here is that anthropomorphism is instantiated in both individual and social biases that follow innately from the human mind's embodiment, the effects of which can transcend both human physicality and the features of human consciousness, known as 'qualia.' It is evident in the fundamental attribution error, for example, in which personality-based explanations for behaviors are overemphasized at the expense of situational explanations (Langdridge & Butt, 2004).

Anthropocentric bias is inescapable in the human condition but is also one of the least well understood innate psychosomatic, existential, and cultural properties that belong to people. It follows from a functionality of their being. There may be adaptive advantages of anthropocentric bias, such as social empathy, and disadvantages of being non-anthropocentric, such as environmental degradation and sociopathy. Adapting from Lakoff and Johnson (1980, p. 17), we need to explore the consequences of ". . . not just the innocuous and obvious claim that we need a body to reason; rather . . . the striking claim that the very structure of reason itself comes from the details of embodiment." Consciousness of embodiment is problematic. Awareness of human selves and others involves anthropocentric predispositions. By contrast, non-anthropocentrism (Panda, 2006) emphasizes the cause of non-human entities, which may be sentient or otherwise, and recognizes their significance regardless of their usefulness to the world of mankind. Non-anthropocentrism therefore involves consideration of the environment apart from the interests of mankind. This becomes especially relevant in current debates of global climate change and environmental ethics, which are concerned with the effects that human beings bring about upon nature, on the balance of our ecosystems, and ultimately on the survival of the earth's biosphere.

The question remains whether non-anthropocentrism or anti-anthropocentrism is in fact achievable at all by humans. Even if people abide by preservationist ethics in their interaction with nature, nature becomes thereby affected with humanistic values. More than this, by extension, the anthropic principle holds that the universe was designed for the perpetuation of a human intelligence to which the natural world is often seen as secondary. To bring the problem a little closer to hand, regardless of whether economic growth is best served through ecological protection, anthropocentrism remains inescapably problematic in so much as it is an intrinsic—although largely unquantified and unqualified property—effect of being human.

Anthropomorphism is implied in the understanding that, as Thomas Nagel (1989, p. 254) has suggested, there is no "view from nowhere." Even the most apparently objective of observers cannot but see through their own eyes. Certain questions arise from this observation: Is it possible therefore for people to decontextualize themselves at all from their environment? How and in what ways might decontextualization be achieved whilst maintaining beliefs that are beneficial to human life and environment?

Artificial intelligence and human cloning are two attempts to distance humanity from existent human ontologies and/or to replicate human functioning in other existential modes. However, it is arguable that in the design of artificial intelligence or in the division of human cloning, humans are merely extending functions of their human selves by a factor of remoteness into their environment, whereby the human form is cast into something less familiar. Human beings cannot but be situated in the contexts and environments in which they function and interact, although the nature of this contextualization is less clear with anthropomorphic robots and clones. It is only in the late twentieth century that space exploration, biomechanical, and virtual technologies have extended the human environment into remoter realms, each of which may be subject to an anthropocentric view.

Human perception is influenced through the psychosomatic feedback loops belonging to the sensory functioning of the perceiver in the environment, as part of an autopoietic system in which the observer creates reality in the act of observation. Such reflexive feedback in living systems follows from a function of their being. This is scientifically observable in the biosensors of medical data, and more remotely in biophysical solutions to engineering and scientific problems such as infrared and thermographic cameras, sonar (sound navigation and ranging), and radiocarbon dating, (a radiometric dating method using the isotope carbon-14 to determine the age of carbonaceous, once living materials up to the age of approximately 60, 000 years).

Classical experimentation in the Newtonian paradigm is influenced in its design by the scientific determinations of the experimenter. The results of any given experiment are also interpreted by a set of any given human pre-dispositions. Likewise, in quantum experimentation (Bohm, 1990), the position of the observer has been shown to actually influence events in the quantum world. As Zabierowski (1988, p. 338) puts it, ". . . the possibility of the physical description of the world is connected inseparably with that of the observer." What these debates and findings point to is that a fundamental anthropomorphic bias can be investigated in which there can be no view that is unaffected by the position of the viewer at the classical Newtonian level, or in the account provided in quantum mechanics. Looking at the mind itself changes the nature of the mind.

Anthropocentrism and Anthropomorphism:
Definitions, Ontologies, Problematics and Reflections

A key problematic concern of anthropocentrism is the assumption that any attempt to explain experience must start from a human perspective. Anthropocentrism and anthropomorphism assume a temporal preoccupation in which human experience arrives to us before experience of all else. There is a temporal ontological priority of beingness that gives humans pre-eminence over their interaction with environment.

As Tom Tyler (2003, pp. 268–269) observes, there are three ways of exercising anthropomorphism conceptually. The first is the literal practice of attributing physical form to a non-human entity; the second is attributing distinctly human qualities to real or imaginary creatures, and the third applies to science in attributing mental states, including intentionality (thoughts, beliefs or feelings), purpose, or volition to creatures that we assume do not have these states. For animal behaviorists and evolutionary theorists, the terminology of anthropocentrism and anthropomorphism demarcates a particular aspect of a species narcissism that has been vitalistically or teleologically working toward some ideal type or interpretation. In this sense, it shares conceptual ground with evolutionary biology: an incremental process that is distributed across all populations at all times.

Like evolutionary directiveness, anthropocentrism suggests a quality that is difficult to isolate, for it is difficult to see mankind's evolutionary development separately from within that development. The conscious striving of humankind to interpret and replicate its experience and to assess future probabilities is dependent on perspectives within current cultures and circumstances. Thus anthropomorphism shares more conceptual ground with evolutionary biology: it is an inescapable quality of our being.

Embodiment biases humankind's view of human society toward the individual's own perception, which may itself be an abstraction of the societal whole. There is no conscious state of pure perceptual experience that is uninfluenced by a person's recognition of experiencing, interpreting, or perceiving independently of that experience. That is, we need language and concepts in which to fit experience. If anthropomorphism is an inescapable fact or effect of life for the embodied human being, then in what ways might an anthropic view be influencing our modes of thought, especially the belief that a classical Newtonian world is accessible to us in an objective form?

Kennedy (1992) casts himself a neobehaviorist (learning theories and models based on the belief that changes in behavior are the result of environmental influences acting with innate predispositions). He argues that the term anthropomorphism confuses function with causes, which is a fatal mistake for scientific inquiry, but he agrees with the claim that anthropomorphic thinking is built into us. Kennedy (1992, pp. 9, 158, 199) and Budiansky (1998, pp. 25–36) argue that mock anthropomorphism is a useful explanatory metaphorical mode, which, while it confuses teleology (the purpose of things) with explanation (the idea of how things work), it nevertheless provides a useful shortcut to meaning in everyday conversation.

Tamir and Zohar (1991, p. 57) argue that there appear to be two kinds of relationships engendered between anthropomorphic and teleological reasoning. In the first, non-teleological reasoning is combined with total rejection of anthropomorphism regarding non-humans. Non-human forms are regarded as having less or more obscure purpose than human forms. Here, functionality must too be ascribed as a property of human reasoning before it can be attributed to others, including non-humans. In the second relationship, teleological reasoning is independent of anthropomorphism in the sense that it does not attribute consciousness to goal-directed behavior. Instead it is based on the assumption that biological systems are structured functionally so that they are adapted to the needs of individual organisms. In this sense, teleological reasoning is associated with overarching biological themes, such as complementarity between structure and function.

Tamir and Zohar, (1991, p. 65) also point out that a non-human entity cannot necessarily be described as functioning with an awareness of its own volition. Furthermore, they establish that most high school students can distinguish between anthropomorphic formulations and factual explanations. They also claim that adults can more readily distinguish between anthropomorphic expressions than can children, arguing that many of them support anthropomorphic formulations as a means of making concepts and processes more comprehensible. Tamir and Zohar (1991, p. 66) also argue for the existence of a clear developmental sequence regarding anthropomorphic reasoning that superficially corresponds to the vitalistic ideas of Aristotle, especially with the belief that life is associated with some unique force that does not exist apart from it.

In some of its popular usages, anthropomorphism implies that human traits are being attributed to creatures to which they do not belong. These traits often take the form of a belief in shared communication between humans and animals (and possibly humans and machines), and the ascription of intentional states to a non-human entity. Heidegger, however, argues that hands and language serve to differentiate mankind from other creatures, thus preserving a hierarchy in which selected features of anthropomorphism prefigures the human. Heidegger's (1996) concept of dasein provides us with a recognition of one aspect of the anthropomorphic problem:

> Attunement discloses Da-sein (or being in the world) not only in its thrownness and dependence on the world already disclosed with its being, it is itself the existential kind of being in which it is continually surrendered to the 'world' and lets itself be concerned by it in such a way that it somehow evades itself. The existential constitution of this evasion becomes clear in the phenomenon of entanglement. (p. 131)

Heidegger's concept of attunement and dasein highlights the embedded nature of anthropomorphic bias as both grounded in human beings and as an existential effect that follows from this. While this quotation points ahead to arguments made in a later section of this chapter concerning quantum physics and anthropomorphism, Heidegger conceives of the concept of consciousness as 'being' apparent to ourselves only in terms of mentalistic evasion and entanglement.

The implication is that existence does not preserve for itself a pure objectivity of recognition. To further highlight the residual anthropomorphism that haunts Heidegger's text (1996), he later states,

> The 'before' and the 'ahead of' indicate the future that first makes possible in general the fact that Da-sein can be in such a way that it is concerned *about* its potentiality-of-being. The self-project grounded in the 'for the sake of itself' in the future is an essential quality of *existentiality*. (p. 301)

This existential property ascribed to being has the status of potentiality. Therefore Heidegger's account describes an innate predisposition that is grounded in the physical domain but that underscores the psychological.

To return briefly to its everyday interpretations, anthropomorphism implies a comparison that is often made in terms of the degree of commonality between humans and non-human entities, which is a projection inappropriate to a particular analytic enterprise. It thereby comes to be seen as an obstacle that obscures the practice of good science. There are two senses in which anthropomorphism is understood to be misleadingly applied. Firstly, it is misapplied in the over-attribution of human qualities to non-human entities. This demeans humans by failing to appreciate their unique traits and by misrepresenting what is distinctive and pre-eminent to humanity. Secondly, by concentrating on what the non-human animal shares with the human one, there is a danger of missing all that is peculiar and proper to the animal as the term may imply non-discrimination of the animal's unique traits, (e.g. the facility of sonar in bats and dolphins).

It is also an error to categorize non-human abilities in terms of human accomplishments. These non-human abilities may require different kinds of thinking altogether. To further explore these common definitions, Caporael (1986) suggests that anthropomorphism may be a kind of default schema applied to non-social objects. Watt (1998) describes the term as bound up in our thinking, which inscribes intentional states to non-human entities. For Eddy, Gallup and Povineli (1993), anthropomorphism delineates the attribution of human qualities to animals according to the degree of similarity between species and mankind and the degree to which an attachment or bond may be formed. Eddy et al. (1993) acknowledge that the term was more prevalent prior to Darwin and there have been successive attempts to dismiss or counter it following evolutionary theory, the behaviorist movement, and the functional approach to the mind advocated by John Searle in *The Rediscovery of the Mind* (1992).

Whilst Watt (1998) argues that anthropomorphism may be endemic to humans, and Eddy et al. (1993) argue that it is 'almost irresistible,' Kremenstov and Todes (1991) state that the long history of anthropomorphism in scientific discussion is inevitable. However, Donald Griffin (1978), the discoverer of bat sonar, emphasizes that the complexity of animal behavior and communication implies conscious beliefs and desires. Griffin argues that the anthropomorphic explanation (the degree of likeness to the conscious experience of mankind) is more favorable than an explanation of animal behavior according only to complex behavioral laws that ignore ethical and social relationships.

From a different point of view, Jeffrey J. Morgan (1995) argues that appeals to mentalistic intentional states are more explanatorily useful than reductively behavioristic rules of unimaginable complexity, which attempt to account for the same states. Morgan argues that attempting to design human thought processes by computer in contexts in which they are unwarranted is just as dangerous as assigning anthropomorphic readings inappropriately, presumably, among other reasons, because it is difficult or impossible to encode human values.

John S. Kennedy (1992) argues that a distinction needs to be made between naïve anthropomorphism of the kind in which, for example, one has a conversation with a dog intimating human-like companionship, and the idea of critical anthropomorphism, which represents a conscious move to push the research agenda of human sciences forward. Such critical anthropomorphism is formulated from data gathered from all manner of sources, such as prior experience, anecdotes, and insights. Rivas and Burghadt (2002) argue that scientific literature is plagued by anthropomorphism by omission, (a lack of acknowledgment that animal subjects perceive the world differently than humans do during the course of scientific experimentation, study, or analysis—perhaps resulting in false, anthropocentric conclusions). They suggest this anthropomorphic bias is especially found in the areas of behavioral ecology, theoretical ecology, issues in zoo management, and decision-making in conservation. Anthropic concerns may also be relevant to city planning, industrial design, and the topics and ontologies of civic research agendas.

Anthropomorphism stresses the human-like characteristics of animals, the animal-like characteristics of ourselves, or the human-like characteristics of objects. Kennedy (1992) suggests that in discussing anthropomorphism, one must also be careful to avoid the nominal fallacy (the belief that giving a name to something is the same as explaining it). Watt (1998) argues for a kind of projective anthropomorphism in which external entities become chimerae through the supposition of intentional characteristics of the observer, similar to the Freudian strategy of assimilative projection. Anthropomorphism is also employed in modeling strategies of human-computational interaction research, which attempts to use computer software to rationalize behavior in human-like analogies. From the perspective of computer interaction research, anthropomorphism is used to augment the functionality and behavioral characteristics of a figure or program in order to relate to it with greater ease.

Some do not believe the anthropomorphic bias to be without benefit. Burghardt (1985) argues that anthropomorphism is a pragmatic strategy to put humans in the place of animals. Such critical anthropomorphism is formulated from data gathered from all manner of sources: such as prior experience, anecdotes, and insights. Primatologist de Waal (2001) called anthropomorphism a heuristic device, useful to study. Daniel Dennett (1989) has incorporated anthropomorphism into his argument for human reasoning involving an intentional stance. However, Dennett's notion of the intentional stance can be seen as non-anthropomorphic in so far as it admits a will to view ourselves and others as 'human like' by virtue of adopting such a stance.

Dennett's argument raises the question of whether or not there actually is a 'construction' of human form being imposed by humans on others (principally on other humans). The argument is that the human form (as conscious and moral) comes to be created through the imposition (or will) of this adaptive strategy. Thus, the intentional stance allows the thinker to conceptualize the possibilities inherent in a deferment of the human form. The existentialism in Dennett's view is apparent in that the imposition of the adaptive strategy of the 'human form' arises out of evolutionary processes which predate it. Therefore, there is a subtle difference between anthropomorphism and the intentional stance, whereas one seeks to describe an effect of human evolution and the other critiques it in the continuous act of 'becoming.'

For Kreuger (2001, p. 19), even the classical scientific methods can be skewed by an anthropomorphic bias. He states that "the selection of hypotheses, their number, their location on the continuum of possible hypotheses, and their prior probabilities depend on the researcher's experience, their theoretical frame of mind, and the state of the field at the time of the study." Kreuger thus suggests that the parameters of experimentation in the classical world are frequently biased by prior states of knowledge on the behalf of the experimenter, even if the experiment itself appears to be free from prior influence.

Similarly, interpretation of the results of an experiment is also prone to bias. From this perspective, for McNeill (1993, p. 25) "it is interpretation or logos itself not anthropos which is the centre [sic] and measure of all things." Humans are not anthropocentric by hierarchy but by preeminence, which returns us to the residual anthropomorphism in Heidegger's analysis in *Being and Time*. For Heidegger (1984, p. 95), human beings are "cornered in the blind alley of their own humanity." Unlike McNeill, Heidegger does believe there is a hierarchy in the biosphere and that humans are at the summit. Human self-consciousness has erected a barrier between the human and non-human.

Bataille (1989), like Heidegger finds that we are condemned to see the world only as humans can: anthropocentrism is the starting point and end result of reflection. It is relevant therefore to look at the criterion by which humans make judgments about their environment. Moral valuation is linked to the aesthetic experience of natural objects, as well as to their utilitarian worth or unique contribution to the environment, independent of human society. From a phenomenological as well as ecological point of view, human beings are at a temporal intersection in their engagement with the earth's biosphere. As Lee points out (2005, p. 236), people are currently constrained by the possibility of having the irretrievable loss of opportunities to have aesthetic experiences of nature. Thus they are faced by the need to act to ensure that future conditions make such opportunities possible. Many environmentalists believe that human beings must adjust their attitudes and behavior to act now or face a virtually barren environment within a century.

As Lee remarks (2005, p. 236), ". . . the conditions that make aesthetic experience possible are precisely those that condition human as well as nonhuman existence...." When people use categories of scientific assignment where no judgments apply to nature, such categories are not coeval with the objects categorized. Similarly, there are no artistic intentions used to judge such categories on the grounds of positive aesthetics. Consequently, the positive and negative aesthetic appeal of nature is beyond our control under conditions of scientific assignment, but not of course the power to alter nature. Frequently then, deep anthropocentrism does not exercise a critical disregard for nature outside of humankind, but rather a predisposition toward logocentrism with regard to nature; humankind is thereby narcissistically blinded to the natural world. Lee's concern is to arrive at an aesthetic criteria relevant to the human experience of nature. She identifies three significant steps to achieve this aim (2005, p. 249): Firstly, we can move away from dualistic thinking of 'subject' and 'object' and acknowledge an interdependence with the environment. Secondly, we must acknowledge that the division between natural and man-made is not absolute and that the two categories are connected through human use and experience. Related to this step is the idea that further analysis of aesthetic experience involves analysis of the conditions under which such categories are made possible. Thirdly, recognition that ideas of negative and positive aesthetic experience of the environment are of limited consideration from an instrumentalist viewpoint, and humanist qualities expressed as a relationship with the environment thereby tend to be overlooked.

Consequently, knowledge can enhance human experience, yet is not a prerequisite to experience over perceptual interest. There is a dilemma here in that something can be correct ontologically yet be insufficiently known; (this is a familiar quality of our dealings with animals). However, something can be indigenously anthropocentric in that it can be molded by perceptual cognition. After all, somatic situatedness is what defines our specific and evolving species membership. As such, we need to recognize that in order to alleviate the effects of anthropocentrism, we have an interdependent relationship with non-human nature that is frequently rendered secondary to our self-awareness and may seem counter-intuitive in the face of perceived threat or instrumentalist gain.

Following Lee's (2005, p. 249) argument, the belief in a world that exists for us alone turns out to be more anthropocentrically biased thinking than truth. It involves the idea that every species may have a place along an inter-related evolutionary continuum in which it should be recognized that the world exists for no single species alone. The question then becomes, do we need to limit or increase anthropocentric thinking with regard to our environment? It depends on how we conceive of the nature of the relationship. There is nothing in our quantum make-up that distinguishes us from other animals; the separation of human sentience and cognition from the ecosystem can no longer be sustained. However, even if there is a repudiation of human-centeredness, it remains present in anthropocentric ways that affect both classical and quantum enquiry, as we are part of the environment which makes the continuum of aesthetic experience and scientific instrumentalism possible.

Can we then modify our environment without also modifying ourselves? What if our anthropocentrism biases our view of nature to our own detriment? These are some of the fundamental ontological questions that contemporary eco-scientific culture now explores. From a political viewpoint, it could be argued that anthropocentrism is a natural bias prevalent in scientific discourse and rooted in a regularly reproduced mind-body dualism: it predisposes the way we think about the human and non-human world toward an anthropological hierarchy and domination in which humans are given pre-eminence. Dualism posits the elevation of the human mind over the body. This is seen as relevant to the way in which human beings dominate and instrumentalize non-human nature and some other human beings.

Plumwood (as cited in Hawkins, 1998, p. 161) argues that dualism has five characteristics: 1) A background of denial, 2) a form of radical exclusion in society or hyper-separation, the division between master and other, 3) an incorporation of relational definition in which the other is defined in terms of lack and only incorporated in society in relation to the master's needs, 4) An instrumentalization or objectification whereby the identity is treated as an object or resource with no volition of its own, and 5) Homogenization or stereotyping, which exercises a uniform membership with no individuality.

In this dualistic scenario, the delineations of a deep anthropocentrism become apparent, which involve a radical separation of humans from nature and potentially even from one another along arbitrary divisions. That such a separation privileges the human is *a priori*. Under these conditions, human epistemologies are indigenously anthropocentric in the sense that they are contextually constrained by our existential being. In one scenario it may require radically different orders of thought for humans to see themselves and a society without anthropocentric bias. However, it may not necessarily mean huge adjustments to take the bias into account when thinking about self and the environment. Rather, a continuous series of smaller adjustments may be required.

The question then needs to be asked: Does this bias have positive or negative effects over the long term or short term? Are there positive biases that follow from anthropocentrism to the survival of all humans in the short term, in inter-special empathy for example, but negative consequences to the survival of the non-human environment and thus the human species in the long term? One solution that sidesteps or ignores the problem is to continually create technological substitutes to the natural world. In this scenario, the reduction of the natural world is not seen as problematic as long as a self-sustaining artificial environment can be built to replace it. This, however, would perpetuate the bias ad infinitum, as well as exhibit deep anthropocentrism with regard to the environment.

Donna Haraway (2001) argues that humans are animals whose claims, valuations, and actions cannot fail to be informed by a myriad of factors that 'fetishizes' existential conditions, for example, perceptual apparitions, culture, and gendered experiences. Following Haraway (2001), Lee (2005, p. 245) suggests that situated knowledge is not necessarily anthropocentric in the sense that objectivity is regarded as a special prerogative of some Western cultural influences, yet she argues that one should avoid poles of Western objectivity and radical relativism. This involves rejecting the binary oppositions and dualisms that characterize false dichotomies. Thus the fast distinctions between the subjective and the objective break down and the alternative is to be concisely anthropocentric, to take seriously the claim that: I begin, I (we) know, in this way (Haraway, 2001). While Haraway's claims are largely from an environmental perspective, from a psychological perspective the anthropic view has not been rigorously studied. Only a few scientists see anthropomorphism as worthy of study in its own right (for example Caporael, 1986; Eddy et al, 1993; Tamir & Zohar, 1991).

Anthropomorphism is considered a hindrance when considered in context of scientific observation, rather than being studied more objectively and taken advantage of. Caporael (1986) argues that if we are unable to remove anthropomorphism from science, then we need to be aware of its presence in scientific assessment. This is a matter of assessing the anthropomorphic bias, as for example Thompson and Barton (1994) have done as an attitudinal measure for given social-scientific experiments. There also exists the need to accommodate the concept of anthropomorphism into thinking about the design of experiments and to devise strategies for recognizing and neutralizing the bias.

From a humanistic discursive viewpoint, it is ironic that religion (and belief in a mind outside the self) is one such strategy, but this is too often inverted for anthropomorphic reasons. Guthrie (1993) has posited that all religions originate in the anthropomorphic tendency to over-detect the presence of other humans and thus the self in the natural world. Milloy (2001) argues that while we may fallaciously ascribe human characteristics to non-human entities, anthropomorphism also emerges when the taxonomic legitimacy of the classification of human is under threat, thereby naturally preserving the likeness of species to one another.

Kennedy (1992) suggests that anthropomorphism acts as a 'drag' on the scientific study of causal mechanism. It does so by influencing cause and effect from the position of the observer, which needs to be taken into account when devising experiments and interpreting data. Anthropomorphic bias has to be factored *in* to scientific calculation, rather than factored out, to gain proper analysis of the components.

Anthropocentrism and Ecology

Among current political arguments concerning global climate change is the claim that the northern elite have an unsustainable mind (Gladwin, Newburry, & Reiskin, 1997). Heath and Gifford (2006) describe this unsustainability as expressed in the unswayable belief in efficiency and economic growth, and optimism about the ability of technology to sustain our environments. As Heath and Gifford point out, appeals to reason alone are insufficient when dealing with anthropocentrism and global climate change. For example, if moral principle often fails to motivate moral action, it is because it inadequately captures a human-centeredness that is informed as much by embodied contingency as by rational consideration.

Heath and Gifford (2006) also argue that capitalism, Judeo-Christian religion and modern technology are inconsistent with environmental preservation. Not only the economic system itself, but the values and beliefs associated with the economic system are also inconsistent with environmental preservation. A form of promotion of profit at the expense of the environment is reinforced by the fact that it has been observed that social cooperation decreases in large-scale dilemmas, implying that the individual's perception that one's cooperation will make a difference decreases in a large group (Hardin, 1968).

Kuhn (2000) argues that when issues of environmental hazard are uncertain, uncertainty is used to justify the discounting of the seriousness of the possible threat. The argument of 'soft' anthropocentrism holds that economic development is by no means pursued at the expense of the environment or interests of future generations, and recognizes that people live in other places at other times. However, this remains an anthropocentric view as the interest in protecting the environment is a human-centered one, apparently non-egocentric but nevertheless configured in instrumentalist terms.

There is a sense in which anthropocentrism and non-anthropocentrism occupy positions on a continuum. Anthropocentrism argues for the separation and consideration of nature from the interests of humans and non-anthropocentrism for the separation of humans from the interests of nature. The former of these positions may lead to a solipsistic technocratic instrumentalism with regard to the natural world and the latter to a position that is regarded as irresponsible in its worship of nature at the expense of the needs of mankind.

Current interpretations of anthropocentric bias are allied to critiques of objectivity, which may obscure the fact that having an anthropocentric bias may come from our deep 'being in the world' (just as a honeybees' 'hymenopteric bias' presumably comes from its being in the world). It is difficult to see how respect for the environment or stewardship (the concept of which may entail limited moral responsibilities) is synonymous with this position, other than to understand that we are co-creators of our habitual contexts. Contrarily, anthropomorphic bias may articulate a caring for the environment which may well need enhancing rather than eliminating. In fact, perhaps the 'ecocentric-anthropocentric' dichotomy is actually a false dichotomy insofar as being human-centric is not necessarily the opposite of being concerned about non-human species or features of the world. It may be possible to hold both positions at different times and in different situations.

Thompson and Barton (1994, p. 149) argue that anthropocentrics may support conservation because quality of life and health may be dependent on a healthy ecosystem. However, ecocentrics judge that nature deserves protection because of its intrinsic value, not only because of its role in enhancing quality of life for humans. Ecocentrics emphasize the connectedness between humans and other aspects of nature that transcends the use of natural resources to satisfy human needs or wants. For ecocentrics, the transcendental aspect of nature is justification for the preservation of nature over and above instrumentalist concerns. Ecocentric individuals will act to support the environment even if these actions involve discomfort, inconvenience, and expense that may reduce their quality of life. The more anthropocentric individuals are, the less likely they are to conserve.

As Thompson and Barton (1994, p. 151) suggest, the anthropocentric view is largely utilitarian; it argues that people are less likely to act to protect the environment if other human-centered values such as material quality of life or the accumulation of wealth conflict with this. They also argue that ecocentrism and anthropocentrism make independent contributions toward explaining apathy toward the environment, conservationist behaviors, and membership of environmental organizations. They also suggest that individuals who support environmental issues for ecocentric reasons may respond to different appeals than those who have more anthropocentric reasons.

Thompson and Barton (1994, p. 155) have devised statistical scales to measure for ecocentrism and anthropocentric attitudes towards environmental issues. Here they posit that individuals may exhibit both behaviors at different times. Thus, programs that attempt to foster interest in supporting environmental action for utilitarian, human comfort, and survival reasons may be counterproductive. A better approach may be to emphasize the intrinsic rewards of being in natural settings through experience in nature and the appreciation of wildlife.

To take up another historical example in current literature, the deforestation of Easter Island as described in Jared Diamond's *Collapse* (2005) presents the argument of the need to urge people to break away from the failures of anthropocentrism. Diamond (2005) suggests that people need to take upon themselves the role of nature's protector and no-longer necessarily see themselves as the center of their environment. However, this raises the question of whether the attempt may simply defer or extend the anthropic bias to a point of further remoteness or perpetuation. Whether it is a successful extension of human values to include nature, or to exhibit a pathology of personification which extends the boundaries of anthropomorphism will depend on the detail and strategy deployed.

Ecocentrics however, might see their own attitudes as a form of holistic caring for the environment that recognizes a care of the human self within it—in contrast to the instrumental separation of humankind from the natural environment. Ecocentricity is thus conceived as something which is separate from a technocratic belief in the possibilities of mankind to synthetically shape the environment, (which assigns no moral value to the natural world apart from the purposes of mankind and thereby devalues the uniqueness of natural systems). However, it can be argued that environmental protection is conducive to producing population quality.

Heath and Gifford (2006) found that perceived knowledge about global climate change was positively associated with the belief in human causes and the belief in the negative consequences of global climate change. They also found it to be negatively correlated with anthropocentrism (that humans are the most important factor) and with environmental apathy. This perceived knowledge was not significantly correlated with ecocentrism (the view that the environment comes first) or with behavioral intention (an inclination to act in a certain way). This indicates a dichotomous reasoning: The anthropocentric view tends to deny that mankind is at the center of global climate change, reflecting an inherent bias that potentially blinds the population to real world effects or to the capacities of individual agency.

Anthropocentrism frequently implies that human kind is the highest purpose of existence in the world; other species are of value only as an extension of the human use of them. Following Heidegger (1996), "being (as *Ereignis*) needs beings so that being may unfold" (p. 31). From the anthropomorphic perspective, human interest is the basis for everything. Protecting human values and rights is the fundamental criterion if human beings are seen to fulfill a progressive role in human history. As de Beistegui (2003) suggests, (following Heidegger), the human is *itself* only by *being* (transitively, creatively grounded in an instituting act). Arguably, this imprint or trace of presence that follows from conscious awareness is what tips the scales in the direction of anthropomorphism, which leads to imbalance and anthropocentric bias in the way we view and manipulate the world—distributed across the population. This is to take in the Heideggerian phenomenological understanding of man to its 'natural' limit. As de Beistegui (2003, p. 277) suggests, "if there remains a trace not of humanism, but of anthropocentrism in the Heideggerian account, it is irreducible".

Deep Anthropocentrism and Counter-Environmentalism

The position of 'deep anthropocentrism' (Jacques, 2006) argues that in contrast to the importance of human development, nature is unimportant. Deep anthropocentrism is the view that humanity is utterly independent of non-human nature, that moral obligation is dependent upon what is relative to humans of immediate benefit, and that there is no obligation for human concern with regard to the environment. This perspective sees humans as fully exempt from ecological principles, influences, and constraints. It argues for a deep division between non-human nature and civilization, and further, that this is inherent to a good society. As Peter Jacques (2006) puts it:

> The stark modern Hobbesian dichotomies that allow for the simultaneous exploitation of both non-human nature and non-dominant human groups institutionalised [sic] by the

> dominant social paradigm between nature and civilization, "savage" and civilized, wild and rational, developed and undeveloped, are fully embodied and strongly held in deep anthropocentrism and the environmental scepticism [sic] which hosts it. (p. 85)

Deep anthropocentrism argues that humanity is the center of concern and analysis, and human welfare is prior to environmental concerns. Nature does play a part in deep anthropocentric conceptualizations, but only in strict instrumentalist terms. Indirect relationships between human welfare and non-human nature, as well as notions of interdependence, are dismissed as soft and invalid. Deep anthropocentrism does not see non-human nature as important in absolute terms, and only in thin instrumental terms is non-human nature considered relatively important. For the deep anthropocentrism, nature, unlike anthropocentric environmentalism, is excised utterly from society. For the deep anthropocentrists, humanity is the center of concern and the analysis of the environment is predicated on human welfare.

The established psycho-economic effects of this can be read either as micro-stages in a larger anthropocentrism or as measurements of inherent inclinations toward bias in human socio-economic relations with the environment. Perhaps it is better to argue for a citizenship in or 'stewardship' rather than dominion and exploitation of nature. However, the moral terrain of this concept is undefined. In this respect, a milder version of anthropocentrism, enlightened anthropocentrism or anthropocentric environmentalism, argues for a concern of mankind with the environment that is not necessarily focused on direct benefits.

If humanity is not interdependent with nature, and humanity has no obligation to nature itself, then human society is released from any expectation or obligation to consequences that may result from changing nature. Consequently, there is no rational ecological citizenship under deep anthropocentrism or environmental skepticism. More importantly, these discarded global environmental problems cannot threaten modernity or its institutions if they are conceptually alienated from society. Such a view also has a moral consequence for people of limited agency in challenging their human rights. As Jacques (2006) has argued, as long as ecological policies cannot show a direct impact of lives saved relative to other mortal threats (such as eating poorly), an ecological criterion falls outside of what is a reasonable and economical course for action.

Anthropomorphism and Human and Animal Differences

According to Bataille (1989), one of the major differences between humans and animals is expressed in the continuity that he sees being shared between animals and the environment. Humans, presumably by comparison, are simultaneously phenomenologically disrupted and enabled by the experience of consciousness and temporality. Memory and Heidegger's isolation of the human feature of hands and the construct of language allow humans to transcend experience in time.

Michael Corballis (2004) has written on the topic of this historical intersection, defined in terms of the moment of 'modernity,' signified by the transition between hand-signed and spoken language. Heidegger also (1984) makes the point that anthropomorphism presupposes that one must assume that one knows 'ahead of time' what human beings are. Any examination of the anthropomorphic effect needs to take into account what it is about human beings in themselves that makes them who they are. Tom Tyler (2003) points out that the term 'humanisation' [sic] by itself does not make sense, for in attributing human qualities to non-humans, we have begun to attribute some quality of being human without specifying what it is.

One common tendency to anthropomorphize is evident in our relationship with animals. For Bataille (1989), there is perfect continuity between the animal and the environment. However, human consciousness is frequently invoked as a measure of human/animal difference, as well as the use of language and the ability to record and manipulate time. Does this mean therefore that the life of the animal is closed to us? Are we condemned forever to our own human and unique perspectives? Heidegger has suggested the temporal preeminence of humans may be involved in anthropomorphism; however, he acknowledges that comprehension depends on awareness and it is difficult to bring out into the open that which is hidden in our being. For Heidegger (1992, p. 84), the hand and language are the distinguishing marks of the human. Ironically, even this conclusion is prone to the bias it describes. However, for Heidegger there is no hierarchy of cause on the animal mind. For him, animals are pure being; their memories are hard-wired as instinct. The hand and its involvement in the development of language points to the difference between humanity and animality.

Derrida (2002, p. 408) has questioned Heidegger's efforts to mark a limit between a non-human living creature and a human. For Derrida, there is an absence of meaning outside a system of differences. He states that human consciousness is a recursive, fractal process involving the navigation through signs. If meaning is embedded in difference, then the anthropomorphic quest becomes one of attempting to infer meaning in its apparent absence.

Tyler (2003) claims that by employing the term anthropomorphism, one has already adopted a set of unexamined assumptions about human beings. The question then becomes how can we conceive, describe, qualify, or quantify this specifically human (anthropomorphic) quality? As Tyler (2003) points out, it is misleading to suppose that the attribution of behaviors 'belong' to creatures that display them, for they are partly constructed by the onlooker who may project a form of species-bias. Einarsson (1993, pp. 78–79) suggests that questioning the continuum or the divide between humans and animals plays a major part in moralizing the natural world through human metaphors, which he acknowledges is a key rhetorical device in environmental campaigns. Xenophanes (580BCE/2001) argues in *Fragments* that anthropomorphism can complicate simple distinctions between the human and non-human, but in so doing it can help us cope with unfamiliarities.

The recognition that anthropocentrism, and more locally, anthropomorphism, are an inescapable residual of being, a tendency of the senses to register themselves predicated by their use, raises the question of what sort of adjustment is required to think in a non-anthropomorphically biased way. This would need to involve not just thinking as others, but thinking also as others conscious of a non-human world. However, this is to acknowledge that such a conception, which remains mediated by current human perception, may be all but impossible. In this respect, anthropomorphism may be characterized by a kind of species narcissism prioritizing human concerns over those of any other species. As Tyler (2003, p. 277) points out, it tends to obscure the difficult question concerning the concept and essence of human nature, and assumes that it has already been answered.

The argument of convergent evolution suggests that while human beings and animals are not constituted mutually and simultaneously, there is a possibility of recognizing new human-animal-technological constructs. While some have argued that animals experience kinds of consciousness, the argument is largely made that most animals are unable to distinguish the difference between themselves and anything else. While the human idea of the animal is both familiar and unfathomable, to conceive of a world without humanity is to conceive of nothing.

Eddy, Gallup and Povineli (1993), following a survey in which they canvassed the attributions of similarity and cognitive function to various animals, revealed a similar pattern of covariation as emergent. This corresponded approximately to phylogenetic group membership, with additional recognition for pets and primates. Thus, anthropomorphic qualities were assigned to animals according to their genetic similarity to humans. Eddy et al. (1993) found that personal similarity was positively correlated to the phylogenetic relatedness of animals to human beings. The more evolutionarily recent animals are, the more likely they are to have higher cognitive functioning. The shape of generalization gradient in inferring mental status to other creatures is thus thought to follow a classical pattern.

A further claim for anthropomorphism is that it is cross cultural, species typical, and almost irresistible. With the advent of behaviorism, anthropomorphism was seen as something to be avoided in discussions of human and animal behavior. Rivas and Burghardt (2001) argue that the term is useful if it leads to testable hypotheses and Gallup, Marino and Eddy (1997) contend that anthropomorphism is a by-product of a unique form of intraspecific selection that gives rise to self-awareness. According to Gallup et al. (1997, p. 88), anthropomorphism involves some form of introspective modeling capacity. If the organism can conceive of itself and can infer the experiences of others by using own experience as a model, then this implies not only an experiential overlap between people, but also empathy and the attribution of intentional states to others.

Gallup et al. (1997) state that self-recognition in an organism allows that organism to situate itself in time and to reflect on its own mental state. This may also entail functioning in competitive, introspective-based social strategies, such as gratitude, sympathy, empathy, sorrow, intentional deception, and other emotions. Emotional continuity between humans and animals is thus more likely than intellectual continuity. For animals, the world is immediate and immanent. We therefore need to ask if human transcendence is illusory. Does our temporal perception and apparent ability to manipulate time make us distinct from animals? According to Bataille (1989, p.

19), "the animal can be regarded as a subject for which the rest of the world is object." We are therefore justified in asking non-anthropocentrically, are there any other creatures that make inferences about mental states in other creatures? This possibility denotes the concept of zoomorphism.

Nicholas Humphrey (1976) argues that the neural hardware of human beings makes most people natural psychologists. Humphrey argues that in an organism that is capable of inferring the mental experience of co-specifics, such as deception or empathy involved in social strategy, then anthropomorphism may be expressed in the social attributions of cognitive abilities to self and others, such as emotions and communicative efficiency. Here, Humphrey acknowledges the problem of anthropomorphic cognition: the attribution of communicative and emotional abilities to creatures that may not feel such qualities. In this case, we need to be mindful of the false consensus effect, which involves potentially overestimating the tendency of others to feel as we do and to share our beliefs.

According to Rivas and Burghardt (2001), Daniel Dennett's model of intentionality goes some way to identifying the nature of anthropomorphic thinking. Dennett identifies four levels of intentional order in cognition: The first order is 'zero-order' intentionality, which attributes enough intentional thinking for minimal mental functioning. Then follows 'first-order' intentionality, an example of which is subjective feelings projected onto another organism. Next is the second intentional order, which may be instantiated when an organism wants another entity to believe that he or she wants something. The third order may involve still more complex intentionality, for example, when someone fears that another will discover that he or she wants him or her to believe something. Thus, anthropomorphism in cognitive states involves the attribution of mental processes that are not only reflexive in nature, but are also equal to or greater than second-order projected intentionality. Rivas and Burghardt (2001) argue that animals do not engage in intentional behavior higher than second-order intentionality.

Anthropomorphism and Quantum Physics

There are at least three possible ways in which anthropomorphism is evident in the quantum physics of the mind. The first begins with the fundamental understanding that quantum physics offers proof that the world we are part of, the world we observe, and our position as observers, brings the world into being. The second way anthropomorphism emerges in quantum theory is as a product of the holism of the 'Many Minds' theory. The third is evident in the theoretical speculation that anthropomorphism can be quantified in the 'white noise' effect of the modified Schrödinger equation (the fundamental equation of wave mechanics which relates wave formation to the allowed energies of wave function).

At the quantum level, the world and we are made of the same stuff. We are constituted in, and part creators of, the world we live in. While in quantum science it is clear that the world observed is in part created by the observer, it is not clear to what degree the observed world is dependent on the unique biological identity of the observer, or rather dependent on their classical position as an observer. Peter Jackson (2002) offers a useful synopsis of the role of consciousness in quantum theory:

> In the transition from the probabilistic quantum realm to the classical realm, a fundamental change occurs, and that appears to be brought about by the experience of the observer. This change takes the technical name of *decoherence*, in which the probabilities described by the wave function collapse to certainty (100% prob). In their unmeasured superimposed state, there are only probabilities, no actualities. But, as soon as we make a measurement, we create a certainty. (p. 7)

The question then becomes: Is it consciousness that brings about the collapse of the wave function in quantum physics? Wigner (as cited in Esfeld, 1999) claims that the content of consciousness is the fundamental reality and it cannot be denied for the individual. The reality of physical objects is, however, relative to their constitution in consciousness. This is in accordance with Heidegger's view of dasein, but not with the arguments of internalism or direct realism (a theory of perception, which argues that we have direct awareness of the external world through our senses). This is in contrast to indirect realism and representationalism, which posits that we are directly aware of only our internal representations of the external world.

Wigner's view (as cited in Esfeld, 1999) is that the existence of physical objects is useful to make sense of the content of consciousness. The content of consciousness is only accessible to the individual; therefore, other individuals are constitutionally equivalent to physical objects. Yet we know from empathy, human emotions, and intraspecific communication that people seem to be more than physical objects — we are predisposed to view each other with an anthropomorphic bias. Thus, following Wigner's view, we are physically separate entities with anthropomorphic features. Our embodied cognitions are independent yet have emergent physical and symbolic qualities.

As Esfeld (1999) points out, in quantum physics, experiments concerning the collapse of wave function as a result of the interaction between the object and the measuring instrument illustrate that there is entanglement between the object and the instrument. Consequently, the object is not in an *eigenstate* (a quantum state that is left unchanged after observation) of the measured observable in classical reality. The measuring instrument cannot indicate a definite numerical value of an object because absolute determination of the object's position in the quantum wave prevents absolute determination of its momentum, (and vice versa). This is known as the measurement problem.

For Von Neumann (1932/1955) this chain is also conditioned by the observer (cited in Esfeld, 1999, p. 146). The observer's body and brain are entangled with the object and instrument. From the position of an observer, the result is a description of the object according to which the body of the observer, including his or her brain, is entangled, along with both object and measuring instrument (Esfeld, 1999, p. 146). The measurement problem can be formulated as the question of how an eigenstate (the measured observable in classical reality) can occur in this wave of indefinable objects (according to quantum reality) in perpetual momentum. Either way, we need a way of explaining this link between classical and quantum worlds. The Many Worlds view argues that there is a wave function for the whole universe and no measurement problem as the experience of the observer causes a branching into another world. The Many Minds view postulates a decoherence in which one quantum state is revealed in one of many possible minds while the universal wave function carries on evolving (Jackson, 2002).

As Esfeld (1999) points out with the Schrödinger equation, a possible solution is that a state reduction is supposed to occur as an "objective event" in the physical realm "before the von Neumann chain reaches the consciousness of an observer" (p. 149) . In other words, a pre-measurement of quantum entanglement is established between the system and observer, achieving decoherence (the collapsing of the wave function to produce one result), by interaction with the environment. However, the existence and entanglement of the observer changes the observation. It is not considered a useful option to assume that consciousness causes state reductions, yet the quantum state applies to all physical systems — the quantum mechanical physical reality needs to be reconciled with experience at the Newtonian classical level of the world.

It may be that humans (as generally thought of in one of our existential modes to be describable as analogical Newtonian classical measuring instruments) are prone to comprehend scale in terms of state reductions, as the classical realm with no entanglement may be the "only way in which nature" can appear to human observers (Esfeld, 1999, p. 151). Yet from remote experimentation, we know that this view doesn't capture how appearances come into being. The Many Minds theory offers a way out of this impasse that also accounts for anthropomorphic bias. Quantum mechanics without state reductions describes the whole of physical reality by assuming that the observer has many minds, in which the observer abstracts from an entanglement what is objectively present (Jackson, 2002).

Von Neumann has suggested it requires consciousness at the point of measurement to collapse the wave function (as cited in Jackson, 2002, p. 8). Given that the experimenter and that which is measured are all made of quanta, the classical and quantum realm entangle in consciousness. In the Many Minds theory, the process of decoherence does not quantify at that measurement point alone, given that there is no necessary intervention by the consciousness of the observer. In this view, there is no problem of measurement because the experience of the observer does not contradict the quantum states. In one mind or in an infinite number of possible minds, the wave function predicts a yes and no, and all the probabilities in between (Jackson, 2002, p. 8). Neither is this to contradict the role of consciousness in classical experiments where outcomes are not thought to be dependent on the observer. Again, to state a paradox, such outcomes could not be known without the presence of an observer. Perhaps it is better to view this as one form of measurement (classical) working towards the outer limits of the exclusion of consciousness, and the other form of measurement (quantum) to the inner limits of inclusion.

For von Neumann, everything is regarded as being quantum, including the brain of the observer, which corresponds to a mentalistic and positivistic view of reality. Von Neumann found that only consciousness could hold the privileged immaterial position where it is not considered part of physical universe but *res cogitans* (a thinking thing) (Jackson, 2002, p. 8, p. 13). Wigner (1964) argued that the consciousness of the observer led to a collapse of wave function, turning probability into measurement. Bohm (1990) postulates an implicate and explicate order. The former is a substrate for all reality, while the latter is the world of space and time unfolded from implicate order. Acceptance that all is made of quantum stuff does not necessarily entail that consciousness is *res cogitans*, but that it is a different order of thing.

As Jackson (2002) points out, the orthodox view of the probabilities of quantum physics suggests that the electron's indefiniteness is transferred to the measuring apparatus, "but, at the collapse of wave function, the measured state goes into the eigenstate"—which corresponds to the result obtained (p. 14). The Many Minds view also assumes that the entire universe has a quantum state. As Jackson (2002, p. 14) explains, this quantum state is a superposition of states that corresponds to many different macro realms, but where all realms are actual: "The idea is that the world splits at each measurement, like a tree into branches, with 'daughter' worlds for each result."

However, the question becomes if all of the realms are actual, then why can't we see them? The Many Worlds theorists argue that after splitting, these realms have no access one to another. However, the Many Minds theory is not closed to this idea. Anthropomorphism enters the picture because each of the many minds representing different probabilities of the eigenstate may not be entirely closed to one another. There will be probabilistic traces of the other in each, and these traces collectively represent a measure of bias inimitable to the individual's experience — the anthropomorphic trace of the individual for any given state.

The Many Minds theory poses a difficulty for the Cartesian in that there is no sharp distinction between subject and object within the theory. As Bilodeau (1996) reasons, our analytic habits are more to do with how our minds appear to function to us than any necessarily direct natural correspondence. It may be that our notion of the workings of a physical substrate needs to change as we register the shift in our comprehension of our inhabitance within classical and quantum worlds. Yet there are as yet no precise experimental coordinates to the end-point of this objective. Bilodeau (1996) argues that phenomenal consciousness offers an inconsistency in the way we are capable of perceiving our world. However, this dividability into properties and spatial relationships may be entering its final phase. This is known as the hard problem. To transcend this we need a non-classical ontology which is neither physicalism (everything which exists is no more than its physical properties), idealism (the only things knowable are the content of consciousness), or dualism (mental phenomena are non-physical properties of physical substances).

As Bilodeau points out (1996), we cannot necessarily expect that the qualia the mind produces are of the same order as that which produces the mind. There is more to mind's relationship to the quantum world than epiphenomena superimposed on patterns of information processing. Rather, in the Many Minds theory, each possible eigenstate is correlated with at least one mind.

Each mind sees an outcome in the classical world, yet does so containing the possibilities of other minds. Yet as each mind sees an imprint of possibility of the other, distinguishing between minds is not the same as distinguishing between possibilities as there may be many millions of possibilities for any given mental state. Consequently our inhabitancy is probabilistic and the weight of probability entails the anthropological bias.

Squires (1998) has argued that since quantum physical equations do not contain what we observe, they are either wrong or new equations are needed. If we take Squires as correct at the representational level, then non-linear elements need to be added to the Schrödinger equation to account for all the effects of wave function collapse. Because stochastic or non-determined processes are involved in quantum physics, a random white noise process may be identified in the modified Schrödinger equation. This random white noise may theoretically register the imprint of anthropomorphism in the quantum mechanical view; it carries the trace of anthropomorphic bias for each individual eigenstate of many possible minds. As Jackson (2002) points out, "instead of proposing infinity of worlds, we could ascribe every sentient being with a continuous infinity of simultaneous minds, which differentiate over time" (p. 16). In this understanding, one mind per person is expressed as a kind of multi-mind. Thus in the Many Minds theory, anthropomorphism is the cumulative effect of the recognition of one mind to the other. Therefore a glimpse into the field of quantum consciousness offers further evidence for anthropological bias.

Anthropomorphism and Robotics

A further context of anthropomorphizing involves the idea of projective intelligence. This does not concern the question of whether a system is fundamentally intelligent, but rather whether it displays attributes that facilitate or promote people's interpretation of the system as being intelligent and possessing human qualities. Joseph Weizenbaum (1966) created ELIZA, a computer program that parodied a Rogerian therapist. In the 1960s, he convinced many people of ELIZA's intelligence and humanity until thematic redundancy was registered as people found repetition in the program. Furthermore as Kiesler and Gotz (2002) have illustrated, people interacting with robots show strong correlations with the interactions of people making people-people judgments (cited in Duffy, 2003, p. 172). Thus human empathy can be enjoined with non-human entities, provided these entities are in some degree human-like. Anthropomorphizing ascribes non-human intelligence to a form based on observation, which explains behavior in a social environment in human terms. As we have already seen, Daniel Dennett's intentional stance (1989, p. 400) involves consideration of the behavior of an entity by ascribing to it the intentional states (beliefs and desires) of a rational agent.

The anthropomorphism of robotics also raises the question: Could a device be built that is more effective than humans in performing the same functions we do, and would such a device be evolutionary? Shneiderman (1988) implies that people who employ anthropomorphic principles to robotics compromise in design, which leads to issues of unpredictability and spatial vagueness in the modeling of imperfect agent theory. This is seen not as the fault of anthropomorphic features, but as a fault of HCI (human computer interaction), wherein who do designers not attempt to understand people's tendency to anthropomorphize, indiscriminately apply certain anthropomorphic qualities to their design ideas from the perceived expectation of users.

In Nass and Moon's (2000, p. 20) account of experimentation involving computers, individuals "mindlessly apply social rules and expectations to computers," thus replicating anthropomorphic epistemologies in the interpretation of computer functioning. According to Duffy (2002b, p. 4), the stigma of anthropomorphism in the natural sciences is similarly partly based on a reductive rationalization of animal or plant behavior premised on models of human intentionality and behavior. This misappropriation extends to explanations but not necessarily to descriptions of non-human behavior. If the intention of anthropomorphism in robotics is to incorporate the underlying principles and expectations people use in social settings in order to model the social robot's interaction with humans, Duffy (2002b, p. 5) acknowledges that the complexity of such an attempt would involve sets of solutions, not just a single engineering solution.

Duffy (2002b, p. 5) suggests that the role of anthropomorphism in robotics is to take advantage of robotic-human modeling as if it were a mechanism through which social interaction could be facilitated. The aim is not necessarily to build an artificial human, but rather address the question of the threshold of an 'optimal anthropomorphism' (Duffy, 2002a, p. 5). He further argues (Duffy, 2002a, pp. 2, 4), that there are two distinct motivations in building an artificial human. First, the engineering issues of building an artificial entity capable of performing in environments with similar behavioral and cognitive responses as humans, means gaining insight into the way we might rationalize its behavior based on humanlike scenarios. Secondly, building mechanisms whereby computational models can be implemented and tested in order to better understand human beings allows for the development of products or software programs which we may relate to with greater ease.

As Duffy (2002a, p. 2) points out, from a scientific perspective, the use of such terms as "familiar, compelling, natural and intuitive" along with qualia and intentional states, are as difficult to deal with as the notion of anthropomorphism. It is the psychological affiliation with object that presents interesting challenges. Yet what is this but a mirroring of human anthropomorphic attachment to objects onto robots? Anthropology is prevalent in robotics because of a tendency to need such familiarity. Duffy (2002a, p. 4) claims that robotics continues in the project of building a humanoid not because the humanoid is the most efficient design for any given task, but because of an innate tendency to anthropomorphize.

If a near perfect human-looking machine was built, what would be the experience when the human-machine looks back at us? Would this involve either empathy or decontextualization? Would the attribution of intentional states to the entity add a further dimension to the interaction? Would this recognition be called artificial intelligence? Duffy (2002a, p. 2) claims that the perceived notion of consciousness may be artificially attained through such anthropomorphizing. However, the artificial attainment of this intentional state may involve mimicry and be shallow or false from an emotion-based communicative viewpoint.

Gong (as cited in Lahtiranta & Kimppa, 2006) argues that introducing human-like features into ICT (information and communication technology) artifacts ensures better user acceptance and positive user experience in the learning environments in which they are applied. However, he is equally as adamant about the potential risks, negative impact, the creation of unpredictability, and 'vagueness' in the user's response to the anthropomorphized artifact.

Lahtiranta and Kimppa (2006) have also attempted to assess the effect of anthropomorphic modeling on the impact of the patient-physician relationship and the quality of patient care. However, as a consequence of the development of computer/artifact to human modeling relationships, humans may be more easily regarded as objects by other humans rather than as feeling subjects. This notion is balanced against the view that people treat computers and new media as real people and places. While an optimal anthropomorphism would create a balance between positive user experience and complex user expectations, it would also have to mimic the complexities of the human mind, body, and communication, as well as mechanical functioning. Lahtiranta and Kimppa (p. 16) also note that with current levels of technology, it is hard or nearly impossible to mimic the psychological and social behavioral complexity of the human.

The Problem of Anthropomorphism

The above discussion demonstrates that to be human is, more often than not, to see things in a particular paradigmatic way, which biases our view of ourselves in the environment. Furthermore, a biased view of ourselves in the environment affects our co-creation of it. Anthropocentrism is connected to environmental degradation and the possibility for deep biases in experimental design and interpretation. If we are aware of this predisposition to think and behave in ways that may damage our environment but which seem unavoidably pre-determined, anthropomorphic abstraction may work in our favor by providing appropriate environmental contexts in which we might imagine and co-create life. Studying the mind changes the mind. Analysis of bias may lead to behavior and experimental modification. The challenge, having recognized an anthropocentric bias, is to identify and assess ways in which anthropomorphic thinking may benefit us or harm us as a species.

Solutions to Anthropocentric Bias Based on New Attitudes in Scientific and Everyday Behaviors

Anthropocentric bias may be partly overcome by using critical anthropomorphism as an acknowledgement of the place of human centeredness in our social and environmental thinking. Scientifically, we can acknowledge the role that the position of the observer has in the design, outcome, and interpretation of experimental results. Acknowledging the human-centered constraints on research necessitates the observation of a degree of bias. We can measure this anthropological bias statistically with scales such as those devised by Thompson and Barton (1994). A recognition of situatedness may also contribute to the understanding of the anthropomorphic bias. This involves the understanding that an experiment, for example, is not just a system of conditions and outcomes, that there is a performative and ontological component which affects the epistemological outcome. Further, there can be an acknowledgement that disinterestedness and detached contemplation when the subject of study is considered, without the imprint of the projected self, may lead to less self-descriptive experimental outcomes.

Watt (1998) proposes the concept of introjective anthropomorphism in which the observer comes to be, in part, a chimera with the observed system. Watt's idea is distinguishable from anthropomorphism as it involves the modification of a person's behavior to an observed subject (for example meowing to a cat). This is a promising idea, yet it is possible that in this scenario, we may surrender communicative understanding for mere identification. The following 14 points present a counter-attitude towards anthropocentrism. We may resist anthropocentrism:

- By recognizing that each experience is a perspective and not necessarily an end in itself; there is no 'final version' of experience. The Many Minds (Jackson, 2002) theory of quantum science holds that we occupy any one of an infinite number of states for each given moment.
- By recognizing that we are biased to create and perceive in the image of ourselves.
- Through devising scales to measure ecocentric and anthropocentric attitudes towards the environment, such as Thompson and

Barton's Anthropocentric Scales (1994), which assist in scientifically recognizing the anthropocentric problem.

- By promoting scientific experiment that factors in the measure of a degree of anthropomorphic bias.

- By overcoming policy description in which our interdependence with environment is obscured and by promoting policy that recognizes our interdependent relationship with non-human nature.

- By introjective anthropomorphism in which the observer becomes a chimera with the observed (Watt, 1998), necessitating the question: Does this limit or redefine possibilities of communication between human and non-human?

- By assessing ways in which anthropomorphism may be of benefit (for example, through interpersonal empathy or assimilative projection), or of harm (through lack of empathy with the non-human environment).

- By leaving a light footprint and being conservative with earth's resources. It is estimated that the imprint of human habitation on the earth would be erased in 100,000 years if humans simply stopped existing and creating (Holmes, 2006).

- By imagining our societies differently, not in a way in which humanity is threatened by the environment but in a way that acknowledges continuity for future generations.

- By exploring ways of investigation and experimental design that are non-instrumental and not informed by exploitative dislocation or remoteness.

- In acknowledging that there are insuperable difficulties in gaining access to animal 'being,' yet in acknowledging that by rejecting the possibility and value of differences, we may avoid assimilating all experiences of our environment to the range of human possibilities.

- In recognizing that while Charles Darwin's 1859 theory of natural selection sought in part to dispel the anthropocentric view, consciousness of ourselves and others as a species naturally inclines us to preserve an intra-species bias.

- In recognizing that an anthropocentric view inclines us to overdetermine our presence in the environment and at the same time it may undermine our relationship with nature, possibly through an unnecessary Heideggarian death anxiety that pre-eminently separates us from our sense of being.

- In recognizing that mankind does not yet determine what constitutes a being: advent of human 'beingness' lies in the destiny of being itself.

References

Anderson, M. (2003). 'Embodied cognition: A field guide'. *Artificial Intelligence*, *149*, 91–130.

Bataille, G. (1989). *Theory of religion* (R. Hurley, Trans.). New York, NY: Zone Books. (Original work published 1948)

Bilodeau, D. J. (1996). 'Physics, machines, and the hard problem'. *Journal of Consciousness Studies*, *3*(5/6), 386–401.

Bohm, D. (1990). 'A new theory of the relationship of mind and matter'. *Philosophical Psychology*, *3*(2), 271–286.

Budiansky, S. (1998). *If a lion could talk: How animals think*. London, England: Phoenix.

Burghardt, G. M. (1985). 'Animal awareness: Current perceptions and historical perspective'. *American psychologist*, *40*(8), 905–919.

Caporael, L. R. (1986). 'Anthropomorphism and mechanomorphism: Two faces of the human machine'. *Computers in human behavior*, *2*(3), 215–234.

Corballis, M. (2004). 'Origins of modernity: Was autonomous speech the critical factor?' *Psychological Review*, *111*(2), 543–552.

de Beistegui, M. (2003). 'Discussion: Response to Peter Warnek'. *Research in Phenomenology*, *33*, 277–280.

de Waal, F. (2001). *The ape and the sushi master: Cultural reflections by a primatologist*. London, England: Allen Lane.

Dennett, D. C. (1989). *The intentional stance*. Cambridge, MA: MIT Press.

Derrida, J. (2002). 'The animal that therefore I am (more to follow)' (D. Wills, Trans.). *Critical Inquiry*, *28*(2), 367–418.

Diamond, J. (2005). *Collapse: How societies choose to fail or survive*. London, England. Allen Lane.

Diderot, D. (1992). The encyclopedie. In S. Eliot & K. Whitlock (Eds.), *The Enlightenment* (p. 8). Milton Keynes, England: The Open University.

Duffy, B. R. (2002a). 'Anthropomorphism and robotics'. Retrieved from http://www.cs.ucd.ie/cprism/publications/pub2002/AIS B02-Duffy. pdf

Duffy, B. R. (2002b). 'Anthropomorphism and the social robot'. Retrieved from http://www.cs.ucd.ie/csprism/publications/pub 2002/IROS-AnthroSocialRobot.pdf

Duffy, B. R. (2003). 'Anthropomorphism and the Social Robot'. Special Issue on Socially Interactive Robots. *Robotics and Autonomous Systems, 42* (3-4), 170-190.

Eddy, T. J., Gallup, G., Jr., & Povineli, D. J. (1993). 'Attribution of cognitive states to animals: Anthropomorphism in comparative perspective'. *Journal of Social Issues, 49,* 87–101.

Einarsson, N. (1993). All animals are equal but some are cetaceans: Conservation and culture conflict. In K. Milton (Ed.), *Environmentalism: The view from anthropology* (pp. 78–79). London: England: Routledge.

Esfeld, M. (1999). 'Wigner's view of physical reality' [Review of essay]. *Studies in History and Philosophy of Modern Physics, 30B,* 145–154.

Gallup, G., Jr., Marino, L. A., & Eddy, T. J. (1997). 'Anthropomorphism and the evolution of social intelligence: A comparative approach'. In R. W. Mitchell, N. S. Thompson, & H. L. Miles (Eds.), *Anthropomorphism, anecdotes, and animals* (pp. 77–91). Albany, NY: State University of New York Press.

Gladwin, T. N., Newburry, W. E., & Reiskin, E. D. (1997). Why is the northern elite mind biased against community, the environment, and a sustainable future? In M. H. Bazerman & D. M. Messick, (Eds.), *Environment, ethics, and behavior* (pp. 234–274). San Francisco, CA: New Lexington Press.

Griffin, D. R. (1978). 'Prospects for a cognitive ethology'. *Behavioral and Brain Sciences, 1,* 527–538.

Guthrie, S. E. (1993). *Faces in the clouds: A new theory of religion.* New York, NY: Oxford University Press.

Haraway, D. (2001). 'Situated knowledges: The science question in feminism and the privilege of partial perspective'. In M. Lederman & I. Bartsch (Eds.), *The gender and science reader* (pp. 169–188). New York, NY: Routledge.

Hardin, G. (1968). 'The tragedy of the commons'. *Science, 162,* 1243–1248.

Hawkins, R. Z. (1998). 'Ecofeminism and nonhumans: Continuity, difference, dualism, and domination'. *Hypatia, 13*(1), 158–197.

Heath, Y., & Gifford, R. (2006). 'Free-market ideology and environmental degradation: The case of belief in global climate change'. *Environment and Behavior, 38*(1), 48–71.

Heidegger, M. (1984). *Nietzsche volume ii: The eternal recurrence of the same* (D. Farrell Krell, Trans.). New York, NY: Harper San Francisco.

Heidegger, M. (1992). *Parmenides* (A. Schuwer & R. Rojcewicz, Trans). Bloomington, IN: Indiana University Press.

Heidegger, M. (1996). *Being and time: A translation of sein und zeit* (J. Stambaugh, Trans.). New York, NY: State University of New York Press. (Original work published 1953)

Holmes, B. (2006, October 12). 'Imagine earth without people'. *New Scientist, 2573*, 36–41. Available from http://www.newscientist. com/channel/life/mg19225731.100

Humphrey, N. (1976). The social function of intellect. In P. G. Bateson & R. A. Hinde, *Growing points in ethology* (pp. 303–317). Cambridge, England: Cambridge University Press.

Jackson, P. (2002). *Te Whakatu Korero/Working Papers. Quantum physics and human consciousness: The status of the current debate.* Lower Hutt, New Zealand: The Open Polytechnic of New Zealand.

Jacques, P. (2006). 'The rearguard of modernity: Environmental skepticism as a struggle of citizenship'. *Global Environmental Politics, 6*(1), 76–101.

Kennedy, J. S. (1992). *The new anthropomorphism*. Cambridge, England: Cambridge University Press.

Kiesler, S., & Goetz, J. (2002). 'Mental models of robotic assistants'. *Proceedings of the CHI 2002 Conference on Human Factors in Computing Systems*. New York: ACM Press.

Kremenstov, N. L., & Todes, D. P. (1991). 'On metaphors, animals, and us'. *Journal of Social Issues, 56*(1), 81–103.

Kreuger, J. (2001). 'Null hypothesis significance testing: On the survival of a flawed method'. *American Psychologist, 56*, 16–26.

Kuhn, K. M. (2000). 'Message format and audience values: Interactive effects of uncertainty information and environmental attitudes on perceived risk'. *Journal of Environmental Psychology, 20*, 41–51.

Kunda, Z. (1990). 'The case for motivated reasoning'. *Psychological Bulletin, 108*(3), 480–498.

Lahtiranta, J., & Kimppa, K. K. (2006). 'The use of extremely anthropomorphized artefacts in medicine'. *International Review of Information Ethics, 5*, 13–24.

Lakoff, G., & Johnson, M. (1980). *Metaphors we live by*. Chicago, IL: University of Chicago.

Langdridge, D., & Butt, T. (2004). 'The fundamental attribution error: A phenomenological critique'. *British Journal of Social Psychology, 43*, 359.

Lee, W. J. (2005). 'The aesthetic appreciation of nature, scientific objectivity, and the standpoint of the subjugated: Anthropocentrism reimagined'. *Ethics, Place and Environment, 8*(2), 235–250.

McNeill, W. (1993). *Heidegger: Visions, of animals, others, and the divine.* (Unpublished Doctoral Dissertation), University of Warwick, Warwick, England.

Milloy, S. (2001). *Junk science judo: Self-defense against health scares and scams.* Washington, DC: Cato Institute.

Morgan, J. (1995). *Anthropomorphism on Trial.*

Nagel, T. (1989). *The view from nowhere.* New York, NY: Oxford University Press.

Nass, C., & Moon, Y. (2000). 'Machines and mindlessness: Social responses to computers'. *Journal of Social Issues, 56*(1), 81–103.

Panda, R. N. (2006). 'Forster's A Passage to India'. *Explicator, 64*(4), 228–230.

Rivas, J. A., & Burghardt, G. M. (2001). 'Understanding sexual size dimorphism in snakes: Wearing the snake's shoes'. *Animal Behavior, 62*(3), F1–F6.

Rivas, J. A., & Burghardt, G. M. (2002). Crotalomorphism: A metaphor to understand anthropomorphism by omission. In M. Bekoff, A. Colin, & G. M. Burghardt (Eds.), *The cognitive animal: Empirical and theoretical perspectives on animal cognition* (pp. 9–17). Cambridge, MA: MIT Press.

Shneiderman, B. (1988, October). 'A nonanthropomorphic style guide: Overcoming the humpty-dumpty syndrome'. *The Computing Teacher,* 9-10.

Searle, J. R. (1992). *The rediscovery of the mind.* Cambridge, MA: MIT Press.

Squires, E. J. (1998). Why are quantum theorists interested in consciousness? In S. R. Hameroff, A. W. Kaszniak, & A. C. Scott, (Eds.). *Toward a science of consciousness II: The second Tucson discussions and debates* (p. 782), Cambridge, MA: MIT Press.

Tamir, P., & Zohar, A. (1991). 'Anthropomorphism and teleology in reasoning about biological phenomena'. *Science Education, 75*(1), 57–67.

Thompson, S. C., & Barton, M. A. (1994). 'Ecocentric and anthropocentric attitudes toward the environment'. *Journal of Environmental Psychology, 14,* 149–157.

Tyler, T. (2003). 'If horses had hands'. *Society and Animals, 11*(3), 267–281.

von Neumann, J. (1932). *Mathematische Grundlagen der Quantenmechanik.* Springer, Berlin. *English translation in Mathematical Foundations of Quantum Mechanics.* 1955. Princeton University Press, Princeton.

Watt, S. (1998). *Seeing this as people: Anthropomorphism and common-sense psychology.* (Unpublished Doctoral Dissertation), Knowledge Media Institute, The Open University, Milton Keynes, England.

Weizenbaum, J. (1966). 'ELIZA — a computer program for the study of natural language communication between man and machine'. *Communications of the ACM, 9*(1), 35–46.

Wigner, E. P. (1964). 'Two kinds of reality'. *The Monist, 48,* 33–47

Xenophanes of Colophon. (580BCE, 2001). J. H. Lesher (Ed.), *Fragments: A text and translation with a commentary.* Toronto, Canada: University of Toronto Press.

Zabierowski, M. (1988). 'Anthropomorphism and cosmographic principle in the Mandelbrot approach'. *Astrophysics and Space Science, 141,* 333–338.

4

ECONOMIC ENVIRONMENTALISM
Post-Structuralism and Intangibles

Introduction

Making predictions in economic theory is notoriously difficult. Two principle reasons for this are that the market is catallaxic (possesses the qualities of spontaneous order created by exchange specialization) and following Ricardo's law, exponentially complex (with specialization among individuals). Similarly, language semantics is not fixed. As with economics, referring to market trends, interpreting the market, and making predictions about meaning is inherently complex as there is never a single place at which language meaning like the market-economy itself, is 'frozen.'[1] Known values continuously interact with yet-to-be accounted for or even unaccounted for values. Implicit structures of presence, (signifying systems) while long recognized as a feature of symbolic interaction, also have the potential to be accounted for in economics; they can be understood as being implicit in the paradigm shift from structuralism to post-structuralism. Concordantly, this chapter explains the commonalities between post-structuralist theory and intangible capital. The identification of economic intangibles can point to ways in which economic theory (and implicitly, economic prediction) may be enhanced by the interpretative framework of post-structuralism. An economic 'intangible' may be compared to Derrida's notion of the linguistic

---

[1] Under normal conditions, the exception is asset freezes by legal intervention, termed a 'Mareva injunction' in English law.

'trace' (1976). Analysis of intangibles can point to a lacuna in the knowledge of the way value is understood and may in fact be accounted for in the economic understanding of market place.

The Economy as a Semiotic System

The table opposite highlights the role of money as a signifying system in which it is a resource (signifier/signified), an accounting device (signified within syntagatic relations/banks and businesses), a measure of value (indexical sign/banks and businesses), a liquid (icon/digital or metallic currency), or a symbol of power or object of desire (metonym) (Morarasu, 2008, p. 150).

Money as a System of Signs

Forms of comparison between deconstruction, semiotics, and money include, first, that 'money' describes a quality that is both positive and negative (can be added or taken away) and as such it has both deconstructive and constructive agentic properties. The market is a shifting network of mediations, differences, and traces. Due to its ability to be converted into other systems of value, objective, subjective, equivalent, non-equivalent, like language, money is a name for a structural or foundational instability (Bradley, 2008, p. 43). In standard Derridean semiotic theory, a given sound or mark (signifier) relates to a given idea or concept (signified). As such, money may instantiate a 'transcendental signified' – an ultimate presence (premised on a tacit and legal belief), which 'anchors all meaning in itself' – and which replaces the presence of a speaker (Derrida 1976). Semiologists may argue that we are merely trading in one idea of the logos for another as we move through history, i.e., one currency for another, but same logos of exchange value. According to Bradley (2008) the agency of language is bound up in the logocentric story of the sign which equates speech with "pure presence and writing with mediation, difference or absence" (p. 47). Signs are characterized by a state of mediation, making it impossible for them to directly relate to a 'present' meaning or signified. Likewise with physical money or even electronic money, it describes a numerical value but also presents its own a priori 'state of language' as a transcendental signified (Bradley, 2008, p. 47). Similarly, just as the sign cannot be reduced to

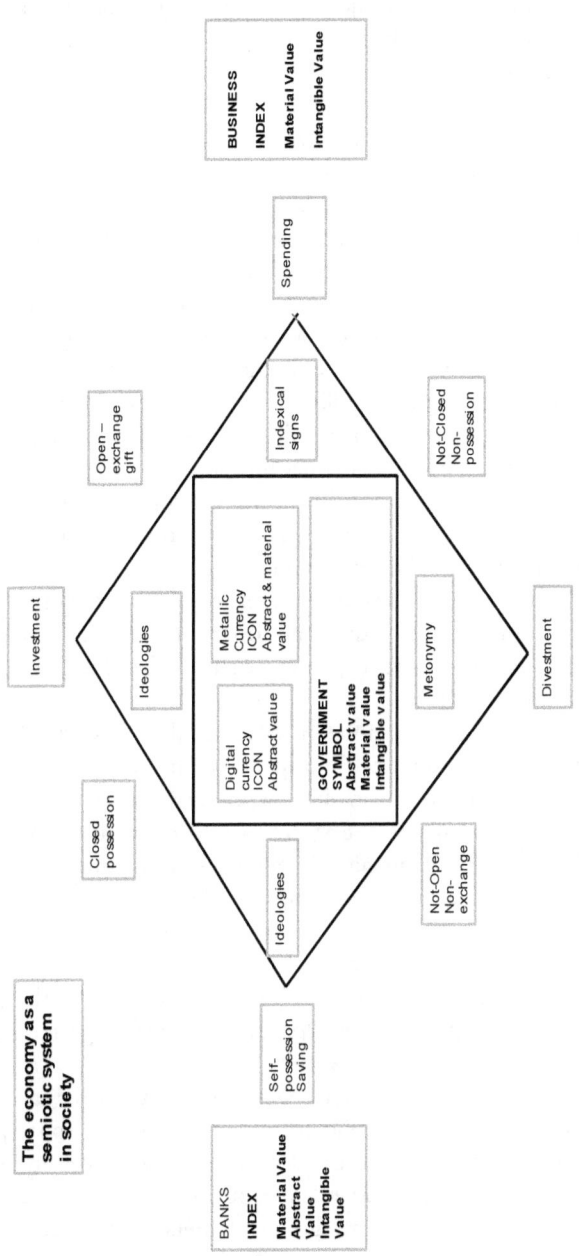

a single presence because it is in a state of complexity, so too with currency, even outdated or extrajudicial currency (because its value as a historical and cultural marker). Nothing is value free and money, even if it cannot be spent, is never value free.

Like money, the sign is first and foremost a mental (or psychic) phenomenon. The idea of money exists independently of physical embodiment as a sign. A digital currency and may or may not be constituted by its material substance – particularly in an intangible form as a transferrable or contractable ration of value. However, there is one important difference in which money differs from the Saussurian notion of the sign, (constituted in the idea that the sign acquires meaning through its difference from other signs) (as cited in Bradley, 2008, p. 66) and that is that money acquires its meaning through its similarity or convertibility to other forms. However, like the sign there is a sense in which the idea of a singular or unique sign is a contradiction in terms. This is because any singular, 'present' or independent sign necessarily contains within it traces of other signs within the system against which it is defined (as does money). Similarly, the currency value of money could be described as a 'signifier of a signified' as every sign works by referring to other signs within the system, but unlike the Sausurian notion, money could be seen as having some inherently positive content of its own: it is not only a symbol for an absent object.

However, money also assumes its place in the network of signs as an instantiation of difference (and to defer (in the sense of delaying something to a later point in time), constantly referring to other elements that exist alongside the system or to elements that "exist before or after it" in the language or value system (Bradley, 2008, p. 71). In fact, money is as much a complex value system as the inter-relation of linguistic signifiers, in which meaning is endlessly transferred or deferred. Indeed, Derrida's notion of Saussure's 'original' trace can be extended to cover not simply "language but the pure realm of thought" a part of the "process of differing/deferring" named as difference (Bradley, 2008, p. 72).

Bishop Nicole Oresma is regarded as one of the original theorists of monetarism in its metallic form. As Galbraith (1992, p. 8) explains, Oresma, "showed how the coinage of gold, silver and copper – coins of fixed weight and reliable purity – replaced the awkward tedium of scales and the weighing of metal." Viewed in this light, it is perhaps ironic that money may be regaining its abstract quality through the digital form in the post-structuralist age. As

Bradley (2008) explains, money as a pure signifier and its financial value or its value as a tradable commodity is similar to the dichotomy between signifier and signified as a portal ". . . through which we can enter the opposition that makes up the metaphysics of presence, such as the opposition between the soul and the body, the idea and the material, the transcendental and the empirical" (p. 72)

What of the intangible quality? For example, in a network of relationships which define the interrelationships of contractual obligations, bank account percentages, interest payable, goodwill, and futures, is the value gained in such transactions mediated through a form of metaphysical system – temporal deferral or other multivariables? Or perhaps the condition of what John Kay calls (2004, p. 100) 'opportunity costs' are at work, defined as the implicit cost involved when the price of doing A making it more difficult to doing B? Both examples invite comparison with the concept of trace as neither sensible nor intelligible. At this point, an economic intangible is neither a subject nor object but a product of the apprehension of a relationship between the two, "neither a property of our being – like being strong or intelligent – nor something that simply happens to us from outside – like capitalism or global warming – but something that exceeds the spatial opposition between interiority and exteriority" (Bradley, 2008, p. 74).

Like the linguistic system of original trace, money is neither empirical nor historic. This state of meaning is even prior to the metaphysical concept because the process of deferral allows them to appear as concepts in the original state – which is the nature of intangibles. The meaning of money as a symbol is a process which is endlessly deferred. Financial value like the 'trace' is an originary condition of mediation, a synthesis of complexity rather than a present being, thing or entity. The trace is nothing: it is not an entity, and it exceeds the question of what is (Bradley, 2008, p. 75). In language, 'the origin' of the metaphysics of presence is never itself present (Bradley, 2008, p. 76). However, if any apparent 'presence' contains traces of what is not simply present in order to be itself in the first place, then this oppositional logic is unsustainable. The interest gained on lending money at a higher rate is a relationship of similarity (like as like), rather than one of duality (being and not-being). Perhaps this quality is what makes prediction in the financial markets so difficult – the financial markets may be run within binary systems but the relations of money and money behavior among people are not of a binary nature.

Economic Intangibles

In a certain world no option would have a value. As Zambon and Marzo (2007) suggest, the combination of uncertainty with flexibility determines the asymmetry of a real options payoff: "at the time the real option could be struck, the probability distribution of value is cut at the level of exercise price; the option is exercised only if profitability limits the losses but not the gains" (p. 33). Can intangibles therefore be considered to be market conditions? Yes and no, a real option for example would be the right to make a value-accretive decision under market conditions. The possibility that the market may add or subtract value from the decision made is the intangible; the fact of its occurrences is the empirical measure, but changes the intangible into something else. (Zambon & Marzo, 2007, p. 7)

Intangible assets generally have six distinct qualities (Zambon & Marzo, 2007, p. 36): The first is that of non-rivalry - an intangible asset can be exploited in multiple activities. The second quality is increasing returns - due to the fact that knowledge is cumulative and benefits are enlarged with its use. The third is company specificity - the value of intangible assets is dependent on the specific qualities of the firm. The fourth quality is path dependency - because intangible assets are grounded in prior decision making. The fifth quality is scarcity - intangible assets are generally not replicable by other companies. The sixth quality is being high-risk - in comparison with tangible assets.

While it has long been recognized that there are at least four types of capital that are non-physical or intangible assets in society, (intellectual capital, social capital, financial capital and cultural capital) and, according to post-structuralist theory, language carries an implied metaphysics of presence, the link between these two concepts, often seen as separate spheres of social scientific discourse, has rarely been posited or explored. Intangible property may also include, a 'thing in action,' such as a debt, a credit balance, the rights inherent in a check, or even rights over land. Linguistic trace and difference/ deferrance (Derrida 1976, Saussure 2002) stemming from post-structuralist critique of structuralism can be compared with economic intangibles in at least two senses. Firstly, they are both concepts of non-physical weightless values. Secondly, they constitute sites of complexity in so far as they apply to either economics as the scientific study of value in society or to the economic relations of language meaning in discourse theory. Derrida's theory of linguistic trace occurs in the context of Western philosophy's pursuit of a met-aphysics of presence As Bradley (2008, p. 6) explains, this is the "... attempt to posit a full or pure 'presence' as the supreme value by which all reality can be judged." Derrida questions whether this view of value based on metaphysical presence is flawed as it can never be accounted for within either a single consciousness or indeed a group of independent conscious minds. Bradley makes a further point that, "[t]he 'here and now' of space and time in which I exist – is actually shot through with an infinite, and almost imperceptible, number of differences, delays or spaces." This is also true of the economics of the marketplace, certainly for the rationale of human decision-making or the myriad of economic choices people make in everyday life. It is in the economic discourse of exchange in society. People make economic decisions based on desires and wants as well as needs. These are often founded on psychological perceptions as well as economic rationalism. Indeed, the entire marketing industry can be said to be based on the communication of intangible *human* wants or desires based on a metaphysics of presence. Intangible capital may be said to exist in copyrights, in patents, and indeed in knowledge itself. The exception to this is 'information' itself. Under English law for example, information cannot be 'stolen' as it is an intangible asset. The English statute dealing with theft was written before the invention of the digital economy, "information cannot constitute property, even though it may have considerable commer-cial or other value" (Baird, Fionda, & Muckham, 2007, p. 246).

Derrida's theory of linguistic trace then adds a complexity to economic theory by "show[ing] that the supposedly primary, domi-nant or superior value implicitly relies on the supposedly secondary, different or inferior value in order to achieve the presence that it should achieve all by itself" (Bradley, 2008, p. 7). Intangible capital is analogous to Derrida's (1976) notion of the mark of the 'absence of a presence', a signifier of 'an always-already absent present.' Corre-spondingly, in terms of tangible and intangible assets, the transfor-mation of money derives from the metallic to the fiduciary, scriptural and digitally coded, which marks a certain de-materialization of money (as a signified and signifier) that has occurred with late-capitalism. De-materialized currencies are coincidental with post-structuralism. As Tratner (2003, p. 792) explains, "Lifting the 'curse' on irredeemable monies is tantamount to lifting the curse on signs that operate without reference, a central element of Derrida's linguis-tic project." Derrida (1976) describes the intangible as the mark of a deletion: "Under its strokes the presence of a transcendental signified is effaced while still remaining legible. Is effaced while still remaining legible, is destroyed while making visible the very idea of the sign" (p. 23). This is also concurrent with the post-Marxist claim that late-capitalism marked a point in history when "production was elevated to total abstraction" resulting in a digital code with an abolished referent (Tratner, 2003, p. 792). While the elevation of capitalism to total abstraction may have been a Marxist observation, it has very real world effects which operate in economic and societal terms, regardless of political perspective.

Simpson (2000) points out that the equations of general equilibrium, which underlie human behavior in the market economy, are too general to be tested directly but give rise to "subsidiary models whose predictive ability are tested regularly" (p. 26). According to the Jevons paradox in economic theory, economic use is equivalent to diminished consumption and new modes of economy lead to an increase of consumption. But what happens if those new modes of economy are yet to be accounted for or exploited? The rationale for a focus on IAs (Intangible assets) follows from the relative inability of economic forecasts to predict turning points in the behavior of the economy (Simpson, 2000, p. 27). While this may be due to the infinite complexity of the interaction between the activities (desires, needs and wants) of people in society, the environment and the economy, it may also be due to the existence of qualities and factors of the marketplace which remain unmeasured. Furthermore, the weightless economy is coincidental with both post-structuralist theory and the measurement of intangibles. As Tratner (2003) explains, "defining a currency in terms of a basket of other currencies does rather seem like a freeplay [sic] of signifiers: searching for the 'meaning' of one monetary sign leads only to an infinite sequence of other signs and ultimately circles back to the same sign" (p. 793).

Although forecasters can predict movements in stable variables with accuracy, they are less successful at predicting volatile variables such as interest rates (Simpson, 2000, p. 28). What determines volatility? Could it not be that the aggregations of unaccounted for measurables play a part in obscuring visible referents of changes and potentials in the market system? According to neoclassical economics, equilibrium theory states that the behavior of supply and demand and prices in the market balances out at equilibrium, in contrast to partial equilibrium. The results of equilibrium theory have meant that faith-based applications of the financial incentive may apply to many market conditions.

Furthermore, intangible assets are in need of definition and understanding as unseen constituents of market volatility (Simpson, 2000, p. 29). This fact raises a series of questions: Does the power of prediction rest upon being informed about present and future prices according to rational expectations and the movement of those prices? What are the limits of rational expectations in the market place given the myriad forms of interactive human psychology and utility? Is it rational to expect to be able to predict when constituent factors of the economy are intangible? Indeed, is prediction possible? Equilibrium theorists might argue that irrational behavior in the financial markets is not possible; this may be understood by adopting the position that the market is an objective entity independent of peoples' processes of rationalization in society. However, the common but imprudent practice in society of buying at high prices and selling at low prices *is* economically irrational (Simpson, 2000, p. 31). If buying and selling (rather than value) are the main determinants of economic activity, does the economy need to be understood as operating rationally, despite the cumulative effects of individual actions that may not be economically rational (but may be psychologically compelling)?

According to rational expectations theory, all investors share rationally formed expectations of future prices and assets (Simpson, 2000, p. 32). But as Adam Smith (1904/1976) famously stated, the chance of gain is often over-valued, and the chance of loss undervalued. Or to take a motivational perspective from social psychology that may reflect a people-centered, Schumacheran (1973/1993) economic view, is purchase based on wants rather than needs in general society itself rational? Isn't the issue to confuse perceptions of psychological value with the market value? Is it possible that constant adjustments between real and perceived value, market value, and the factors of economic growth, may all be predicated on implied perceptions of value—which actually may have a positive and negative valency as well on possible calculation of tangible asset values? We then need to distinguish between the psychological perception of value and the unaccounted for measure of value in an intangible asset. Here we might think of intangibles as closely related to Romer's implication from new growth theory, that they are the latent or endogenous spill-overs from limited monopolies on ideas, or unaccounted for factors of the possibilities of latent growth (1994, p. 13) .

The Characteristics of Intangibles

As John Kay (2004) explains,

> The modern economy has many different kinds of distinctive capabilities and so many different kinds of intangible assets: competitive advantages based on brands or reputations with groups of customers; strategic assets such as patents and copyrights or local monopolies; structures of relationships with suppliers or employees. 'Our people are our greatest asset' is a cliché of company reports, and there is a lot in it. All of these factors explain why the value of companies is greater than the value of their tangible assets. (p. 174)

Intangibles are defined as ". . . assets that one cannot see or touch, such as patents and goodwill, but that become relevant when they are the subjects of a market transaction" (Zambon & Marzo, 2007, p. 51). Zambon and Marzo offer a further definition: "Intangible assets can be defined as a source of future benefits that is without a physical embodiment" (p. 51). For example, intellectual property is an intangible asset with legal rights. This definition includes innovation-related intangibles (research and development patents), but also market-related intangibles (brands), human resource intangibles (competencies and skills, training), and organizational intangibles (internal structures, systems, procedures, routines, and processes). A significant distinction can be drawn between 'hard' intangibles, which are tradable in the market place, and 'soft' intangibles, which cannot be sold or negotiated (Zambon, 2009, p. 2).

Intangibles can also be deployed in multiple issues. An example which Daniel Andriessen (2003) uses is that, "[a]lthough an airplane can be used during a time period on one route only, its reservation system can serve, at the same time, a potentially unlimited number of customers" (p. 5). As well as profiting from network effects, they are frequently characterized by large fixed costs and minimal marginal costs, increasing rather than decreasing returns. Lower prices encourage consumption and discourage increasing production (as high prices encourage people to purchase a commodity, as it is perceived as being of high quality). However, scarcity does not necessarily correspond to price; a commodity can be scarce and worthless.

As Perelman (2006, p. 81) suggests, economists may have confused the continuum between religious, family, and economic values (metaphysical value), but coterminous with this, the market system tends to devastate anything that has a price label, such as air or water. Perelman (2006, p. 82) compares the concept of value to that of gravity, suggesting that the relationships between firms and households hold the economy in balance, just as physical bodies at a distance hold the planets of the solar system in balance. John Stuart Mill (1848,III, 1.1) stated, "Almost every speculation respecting the economical [sic] interests of a society thus constituted, implies some theory of Value: the smallest error on that subject infects with corresponding error all our other conclusions; and anything vague or misty in our conception of it, creates confusion and uncertainty in everything else."

Furthermore, the more permanent an investment is, the more uncertainty pervades the decision making process in its development, but what if some of this uncertainty resulted from an inability to read or measure intangible benefits? Physical theories built around conservation laws, such as the conservation of matter and the conservation of energy, typically are thought to influence economics. However, as a description of economic relations they have very little explanatory power, rather perhaps they can be understood as imitational factors of supply and demand within physical space and time (Mirowski, 1989, p. 271).

Economies operate in something not quite like the Newtonian world because the concept of value can vary wildly. For example economists may use scalar dimensions because in production, the lapse of time is a negative intangible. An economist's understanding of both physics and metaphysics may account for why speculation itself is both rational and irrational, but intangible assets both do and do not operate in a scalar Newtonian world. Such are post-structuralist concepts and values, and the difficulty of their measurement is coterminous but not equal to the abstraction of the weightless digital economy.

Merton and Scholes ("Additional Background," 1997) along with Black won the Nobel Prize in Economics for developing a pioneering formula for the valuation of stock options which signified more efficient management of risk:

$$C = SN(d) - Le^{-rt}N(d - \sigma\sqrt{t})$$

Where variable d is defined by:

$$d = \frac{ln\frac{S}{L} + (r + \frac{\sigma^2}{2})t}{\sigma\sqrt{t}}$$

Their prize-winning formula for valuing complicated assets provides a technique to remove risk from investment, valuing and combining assets to insure against risks. Thus, there *have* been successes in the measurement of intangible effects in the market place. Presumably, the predictive ambit may apply at one time only in a limited area, so may only account partially for catallaxy, or for areas of specialization and their complex relationship with growth (Perelman, 2006, p. 91). The complexity is compounded in the measurement of intangible assets throughout the myriad *potential* intangible accountables of the economy.

'Floating value' is economic value given in the application of regulation, but Perelman (2006) points out there are no discount rates in the natural sciences; while business may value natural resources as no different from paper value, clearly there is a qualitative difference. Money is a manufactured resource, only useful where recognized (Perelman, 2006, p. 93). Economist Arthur Cecil Pigou (as cited in Perelman, 2006, p.97) suggests that people "distribute their resources between the present, the near future, and the remote future on the basis of a wholly irrational preference...but business has no reason at all to consider a lower discount rate for the more distant future." Robbins (1935) observed in an influential study of economic methodology that economy is a "complex of scarcity relationships" (p. 19). Deviation from marginal pricing causes inefficiency and social loss. Furthermore, emotion and intuition exercise more influence in most economic exchanges than abstract or outright knowledge. Most economists might recognize that it can't be assumed that everybody behaves in a rational manner or that everyone's idea of rational behavior is the same.

A significant observation of the relationship between specialization and growth is that specialization forms the basis for high-income levels and productivity. There are disadvantages as well as advantages to specialization. These are described by the *efficiency* effect, the *risk* effect and the *dynamic* effect. A specialized firm is supposed to be able to exploit economies of scale, reap the benefits of learning, and use specialized inputs (efficiency increases). However, specialization is disadvantageous if the firm is locked in a mature, declining industry. Mature industries or those with a low potential for product differentiation will not be able to grow fast (or produce dynamic effect). Furthermore, as Aiginger (2001, p. 11) points out, if member countries of a monetary union are too highly specialized, the external shocks will lead to asymmetries in demand and can't be compensated for by changes in the external value of currencies.

Does growth make structural change necessary? Or vice versa, does growth depend on past structural change more closely than on innovation by itself (Aiginger, 2001, p. 39)? A three-sector hypothesis that was proposed by Fourastie (1954) and Clark (1957) posits a cycle of growth in per capita income in which the share of the primary sector decreases with rising income. The secondary sector of manufacturing, construction, and utilities increases, later losing a proportion of market share in production and demand, but the service sector continuously grows. There is a fourth sector, the information sector, which is spread throughout the various sectors. Innovation is more likely to be found in the second and fourth sectors as these sectors drive the shift of the demand curve (Aiginger, 2001).

Intangible Assets, Real Options and Risk

Berk, Green and Naik (1998, 1999) show that research and development projects and new ventures display high level of systematic risk. Ho, Xu and Yap (2003) empirically demonstrate that research and development investment increases a firm's systematic risk. Wyatt (2002) remarks that the associated risk to intangible assets is higher than associated risk to tangible assets, since generally intangible assets precede investment in tangible assets (Marzo, 2007, p. 36). So investment in intangible assets is characterized by uncertainty. As Marzo points out, risks reduce during the investment phase, are high during the research phase, and reach a lower level at marketing phase. The arrival of new information and knowledge makes the reduction of risk possible during the life of an investment in an intangible asset (Marzo, 2007, p. 36).

It follows then that intangible assets are far from being value-free. Significant growth can arise from investment in intangible assets. Furthermore, risk reduction can itself be an Intangible asset. For any given project, a reduction of the value of the real option is positively correlated to the risk variance of the underlying asset value (Marzo, 2007, p. 43). Despite the fact that the value of intangible assets may be calculated from comparing the market value of stock to the accounting book value, there is no standardized measurement of Intangible assets.

Positive Intangibles	Negative Intangibles	Remediation of Intangible threats
Human capital (knowledge)	Threat of substitute products and services	Environmental scanning
Organizational capital (Collective knowledge, policies, regulations)	New market entrants	Regulation
	Switching costs	Education
Information capital (Intellectual property, patents)	Slack (time lost)	Market capture
	Barrier to entry	Rights
Market Positioning	Obligations	Networking
Statutory based	Complaints	Known unknowns
Customer based	Unknown unknowns	Fulfillment of terms
Market based	Threats to supply	Patents
Contract based	Barriers to research and development	Motivators
Technology based	Product differentiation	
Social capital	Planning	
Bonding capital		
Bridging capital		

So investing in intangible assets reflects investment in market-based uncertainties that may yield higher returns. The amount of risk may reduce during an investment. It is at its maximum at the moment of the conception of the research project, while new information and knowledge make the reduction of risk possible as it reaches a lower level at the moment of marketing phase. As Marzo (2007, p. 43) observes, the value of real option is positively correlated to overall risk (variance) of underlying asset value. Additionally, Marzo (2007, p. 49) suggests that the Generally Accepted Accounting

Principles (GAAP) need to be supplemented by Generally Accepted Intangible Principles or GAIP, but these have yet to be devised.

As Zambon (2009) states, "... even though then relevance of Intangible assets is unanimously recognized, the delicate and complex issue of the valuation of and reporting on intangibles remains wide open" (p.1). So if intangible assets are those that one cannot see or touch but which become relevant when subject to market transaction, does one compute value of intellectual property in terms of items and cost of development, or also on the basis of forecast earnings based on current value? Marzo (2007) recognizes that the calculation of value in intangible assets doesn't include political and economic factors such as those, for example, which influence share prices. Yet there is no standardized measurement or market valuation formula based on imperfect information of the value of Intangible assets.

Conclusion

Economic intangibles are non-rival assets. They are assets that one cannot see or touch, such as patents and goodwill, but become relevant when subject of a market transaction or even a potential transaction. They offer future benefits or detriments without physical embodiment. The rationale for focus on intangible assets follows from relative inability of economic forecast to predict turning points in economic behaviors. If economic use is equivalent to diminished consumption, what if, as well as devaluation, new modes of economy lead to aspects of consumption based on concepts of the 'yet to be seen' or 'exploited?' The point is not simply the transposition of psychological needs, wants, and desires with economic rationale, rather that intangible assets both affect economic behavior and enhance the value of capital assets. The economic measurement of intangibles involves the complex interaction between people, goods, services, aspirations (needs, wants, desires) and qualities of capital which have yet to be measured in generally accepted intangible principles (Marzo 2007). It remains a task for economists and accountants to devise standardized measures of the value of Intangible assets and apply these to models of economic theory and accountancy practices.

References

"Additional background material on the bank of Sweden Prize in Economic Sciences in Memory of Alfred Nobel 1997." Retrieved 30 December 2010 from: http://nobelprize.org/nobel _prizes/economics/laureates/1997/back.html

Aiginger, K. (2001). *Speed of change and growth of manufacturing. Structural change and economic growth: reconsidering the Austrian 'Old-Structures/High-Performance' paradox.* Retrieved from: http://www.oecd.org/dataoecd/24/23/2076797.pdf-

Andriessen, D. (2003). *Making sense of intellectual capital designing a method for the valuation in the intangible economy.* Oxford: Elsevier Butterworth Heinemann.

Baird, N., Fionda, J., & Muckham, M. (2007). *Criminal law.* London: University of London Press.

Berk, J., Green, R. C. and Naik, V. (1998). 'Valuation and Return Dynamics of Research and Development Ventures'. Retrieved from http://papers.ssrn.com/sol3/papers.cfm?abstract_id=469 08.

Berk, J., Green, R. C., & Naik, V. (1999). 'Valuation and Return of New Ventures'. Retrieved from: http://papers.ssrn.com/sol3/papers.cfm?abstract_id=133653.

Bradley, A. (2008). *Derrida's of grammatology: An Edinburgh philosophical guide.* Edinburough: Edinburough University Press.

Clark, C. (1957). *The Conditions of Economic Progress.* London: MacMillan.

De Saussure, F. (2002/2006). *Écrits de linguistique générale* (edition prepared by Simon Bouquet and Rudolf Engler), Paris: Gallimard. English translation: Wrings in General Lingusitics. Oxford: Oxford University Press.

Derrida, J. (1976). *Of grammatology.* (G. C. Spivak, Trans.). Baltimore and London: The Johns Hopkins University Press.

Fourastie, J. (1954). 'Predicting Economic Change Before Our Time'. *Diogenes, 2* (5), 14-38.

Galbraith, J. K. (1992). *A history of Economics: the past and present.* London: Penguin.

Ho, Y. K., Xu, Z., & Yap, C. M. (2004). 'R&D Investment and Systematic Risk'. *Accounting and Finance, 44,* 393-418.

Kay, J. (2004). *The truth about markets – Why some nations are rich but most remain poor.* New York: Harper Collins.

Marzo, G. Intangibles and real options theory: A real measurement alternative. In Zambon, S. & Giuseppe Marzo. (2007). *Visualizing intangibles: Measuring and reporting in the knowledge economy.* Hampshire: Ashgate.

Mirowski, P. (1989). *More heat than light: Economics as social physics, physics as nature's economics.* Cambridge: Cambridge University Press.

Morarasu, N-N. (2008). Managing money and language as interrelated sign systems. Available from *http://ub-ro.academia.edu/Nadia Morarasu/Papers/354312/Managing_Money_and_Language_ as_Inter related_Sign_Systems*

Mill, J. S. (1848/1909). W. J. Ashley (Ed.) *Principles of Political Economy with some of their Applications to Social Philosophy.* (7th Edition). London: Longmans, Green and Co.

Smith, A. (1904/1976) *An inquiry into the nature and causes of the wealth of nations.* Chicago: The University of Chicago Press.

Perelman, M. (2006). *The perverse economy: Scarcity, extraction and value in economic theory.* New York: Palgrave. Macmillan.

Robbins, L. (1935). *An essay on the nature and significance of economic science.* 2nd edition. London: Macmillan.

Romer, P. M. (1994). 'The Origins of Endogenous Growth', *The Journal of Economic Perspectives, 8* (1), 3-22.

Schumacher, E. F.. (1973/1993). *Small is beautiful. A study of economics as if people mattered.* London: Vintage.

Simpson, D. (2000). *Rethinking economic behavior: How the economy really works.* London/New York: St. Martins.

Tratner, M. (2003). 'Derrida's debt to Milton Friedman'. *New Literary History, 34* (4). Pp. 791-806.

Wyatt, S. (2002). 'Accounting for Intangibles: The Great Divide Between Obscurity in Innovation Activities and the Balance Sheet'. *Singapore Economic Review, 42* (1). Pp. 83-117.

Zambon, S., & Marzo, G. (2007). *Visualizing intangibles: Measuring and reporting in the knowledge economy.* Hampshire: Ashgate.

Zambon, S. (2009). *IPR, intangibles & valuation: Visualising information for finance access.* Retrieved 22nd December 2010 from: http://www.wipo.int/edocs/mdocs/sme/en/wipo_smes_rom_ 09/wipo_smes_rom_09_f_workshop_03_1-main1.pdf

5

COMMUNICATION CONTEXTS
Psychologists and Medical Practitioners

In medical psychology, client and practitioner conversations are a node within a network of conversational relations. While the ethics of client confidentiality ensures that these conversations are contained, there is a wider role that medical practitioners may enjoy as practitioners and promoters of public good health. Medical practitioners are also communicators.

Michel Foucault (1988, 2003) argues that historically speaking, madness and reason are on a continuum. He posits almost a dialectical relationship between them as constructs or states of being, defined at once by presence and absence. From the perspective of historical review, Foucault notes a change at the end of the eighteenth century where there is a fissure in communication between reason and unreason. This is characterized by a state in which unreason has been 'delegated' to an authority outside the subject and objectified in the abstract universality of disease, or an equally abstract construct of reason. The continuum between the two was severed and the language of commonality ceased. Foucault (1988, pp. x-xi) states, "The language of psychiatry, which is a monologue of reason about madness, has been established only on the basis of such a silence." Perhaps this echoes the relationship between reason and unreason that pervades modern society also. Kirmayer (2008) makes a more subtle point when he claims that "the moral project of medicine is partially hidden or obscured by the use of rhetoric of science that portrays clinical practice as primarily technical, evidence-based, and neutral or value free" (pp. 127-128). So is medical science or psychological medicine and its practices of communication a 'science' or 'art'? Should we value it in more or less if were either? As Waymack (2009, p. 221) states, "even if and when genetic variability is 'understood' the subjective (and inter-subjective) nature of human persons – biographical, socio-economic, political, spiritual, cultural – prevent medicine from ever truly being a 'science' in the traditional sense." Waymack invites the familiar problem of 'ultimate' objectivity, is it possible for a human to look at human in terms that preserve complete detachment?

As Phillips (2005, p. xiii) states,

> A word with few synonyms, 'sanity' has always been an unfashionable term that has never quite gone out of fashion. First used by physicians in the seventeenth century to refer to 'health in body and mind,' its more familiar modern connotation as the opposite of or antidote to madness only really developed . . . in the nineteenth century.

Clearly our society also is reductive to reason, but for different reasons than those of the pre-enlightenment. As Phillips (2005) suggests, "sanity and soundness . . . are like propaganda for a world, a 'quality or state of being' that has never existed. Sanity and soundness cannot easily coexist with suspicions about their own soundness" (p. 40). Phillips makes the point that human experience is broad and that almost everyone may be challenged at some time in their life by conditions that effect their sense of 'framing' as they make sense of their experiences. He also makes the point that 'sanity' and 'soundness' are descriptors that are most often used extrinsically by people and as such may seem glib in the face of life's complexity, to say something is 'sane' might mean little more than that we understand it. Phillips may be closer to the mark when he claims that we should do away with 'madness' than Foucault's suggestion that psychiatry is based on a language of ceased commonality.

Furthermore, peoples' expressions of their thoughts, beliefs and experiences are culturally situated such that meaning is collectively produced. As Kirmayer (2008) suggests, "the biological effects of diet and lifestyle break down the culture/nature dichotomy, since culturally determined patterns of behavior refashion human biology" (p. 129). Thus people express themselves and are expressed by their environments. Kirmayer (2008) continues by saying that: "On the face of it, mental health problems may reside not only in defects in the hardware of the brain but in its software (learning) and outside the individual in interpersonal interactions in families and communities, as well as in larger social configurations." (p. 131) Although ardent Darwinists would claim that evolution is not possible over the time-scale of a given individual's life, everyday sensory experience can lead to a weakening of some synaptic connections and a strengthening of others. Certainly the 'learning' context aims to strengthen the connections the people make to discovered knowledge and equip to make discoveries themselves. Therefore a factor of contemporary medical understanding of human brain, mind, and behavioral functioning is the idea that mental processes culturally produced as well as being biological and organic. (Kandel, 1998, p. 464)

Why this is relevant to environmental communication is that culture and implicitly the human environment 'produces' behavior in the form of attitudes and dispositions. At one end of the spectrum of ordinary comprehension, the ability to distinguish among its forest of signs entails that meanings may sometimes collapse or implode, a process Baudrillard (as cited in Littlejohn, 2002, p. 308) refers to as *hypertelia*. Foucault (1998) also distinguishes between a 'physical truth' and a 'moral truth,' the former being the accurate relation between our senses and physical objects and the latter being the accuracy of the relations between moral objects or between ourselves and those objects. The Russian psychologist Vygotsky (as cited in Lewis, Pea and Rosen, 2010, p. 353) acknowledged some of this complexity in cross-cultural symbolic interactionism in 1978:

> The mediating signs people use to understand and represent the experiential world, from language to their signifiers, form a generative basis for human psychology and culture. Intra-psychological encounters with the uses and thereby the meanings of such signs give rise to inter-psychological mental structures and processes . . . the tools we build to mediate these symbolic activities change the ways humans think. By building tools, people build the material basis for consciousness, transforming the environments and restructuring the functional systems in which they act and learn.

It may seem astounding to us now that in the classical era psychology (as the social science we understand today) did not exist;

physical therapy could barely be separated from religion. The debate over whether the practice of medicine is more of a science than an art has endured various convolutions since the seventeenth century but views have progressively shifted in the direction of science.

So perhaps there is a fundamental change between viewing medical society in the nineteenth century as normalized around scientific rather than religious principles, and medicine regulated in accordance with disease – one in which a physician was him or herself of normalized on the periphery of a system of symptoms. The individuation of the twentieth century society saw the disambiguation of notions of non-secular suffering. The growth of the modern hospital may have led to this reversal. The immediacy of the relationship between sickness and alleviation, articulated into analyzable elements, represents a change in the medical experience and the rubric of how the understanding of disorder is formalized. Disease would no longer be a sign of the medieval God's punishment. Rather, the detachment from religion would signify not only God's indifference but also the inaptitude of this paradigm of understanding to offer a cure for life's ailments.

Psychotherapy

Psychotherapy is seen as one common route out of internal complexity or from feelings of irrationality that cannot be governed by our normal ways of making sense. Psychotherapy extended the conversation of what is normal and abnormal into the realm of scientific understanding rather than religious or existential dilemma. Psychotherapeutic understanding is premised on increasing individuated knowledge of the range of behaviors that cause problems and provides strategies for overcoming them. Phillips (2002) explains that,

> It was part of the initial exhilaration of psychoanalysis – the thing that it made it seem to some people both a liberation and an inspiration – that it seemed to extend the permissible in thought and feeling. That by acknowledging the sheer morally ambiguous complexity of what people might have to say – and acknowledging the very real difficulties involved in saying it – new kinds of life seemed at once possible and plausible." (p. 46)

Psychotherapy, the place where people unlearn bad cognitive and behavioral habits and re-learn good habits, is of course the clinical branch of discursive psychology, known as the 'talking cure'. One of its primary functions is as a defense against anxiety, manifested in myriad presentations of symptoms in response to the pressure of modern life. This involves the presentation of a self (which may be tentative and contradictory) in front of an interlocutor. Kirmayer (2006) defines it thus, "Discursive psychology aims to identify distinctive modes and practices or representing self and other that are constitutive both of personality and illness explanations" (p. 134). Consequently, the link to Foucault is found in the attempt of psychotherapeutic practices to base symtometology on a specific frame of reference which utilizes an index of distress. Psychotherapy also accentuates the fact clinical treatment is inherently based on language and narrative interventions. Even those who seek psychotherapy do so in the hope of learning something both subjectively and objectively useful about themselves and their lives. Phillips (2005) states that,

> The possibility of there being at least sane versions of ourselves has traditionally been the source of our hopes for ourselves; and the idea of sanity, in which we have invested so much, has always been something that we could, at least, look forward to. (p. xviii)

Medical client and practitioner communication

As Perkins and Sanson-Fisher (1996) state, there are two domains in quality medical care, technical competence and effective interactional skills. Assurance in the former comes more easily than the latter. The importance is underscored by the fact that the medical interview is the basis for 60% of medical diagnosis and treatment decisions (Perkins & Sanson-Fisher, 1996, p. 17). Furthermore skills can be divided into three kinds (Perkins & Sanson-Fisher, 1996, pp. 17-18):

1. General communications skills that gather information.
2. Interpersonal communication skills, including proxemics, conduct, control of item, empathy, and warmth.
3. Information transfer skills that are involved in increases in the probability of outcomes.

The use of skills is more likely to result in problem detection and to assist in patient recall—which facilitates compliance. Communication problems are the result of the interactions between doctor and patient. These interactions are more than one-sided exchanges: the doctor volunteers knowledge, the patient helps others by providing knowledge about problems that patients might be afflicted with. The doctor may either objectify the patient to cure the illness or objectify the illness to treat the patient. The patient's condition may prevent him or her from understanding the doctor, or the doctor may be reluctant to follow the patient's suggestions and not enquire beyond the initial problem.

Evidence has shown that some poor hypothesis generation and problem solving may result from a 'high control style' and a premature focus on medical style, which can result in an overly narrow approach to hypothesis generation. It may prevent the patient from articulating all of his or her concerns. It is also rare that doctors will ask patients to volunteer their ideas. The latter is particularly complicated during the psychological interview in which the doctor will be assessing the patient on the nature and quality of his or her thought, not just on the context that he or she might think immediately concerns its expression. However, research by Waitzkin (1984) suggested that in 65% of encounters, physicians underestimate a patient's desire for information. For many patients, aside from the cure or clinical management of their condition, they seek affirmation from their physician. Not only is there a need for information but also the information needs to be given in the context of the patient's understanding or potential to understand.

Perhaps this is due to three reasons. There is disagreement about the importance of relating different types of medical information (patients generally are more concerned with prognosis, diagnosis, and causation than simply information about their condition). There is also disagreement about the value of giving information, (uncertainty regarding future support). Thirdly, there are problems with patient recall. Open rather than closed questions and attentive listening may aid patient understanding and facilitate diagnosis. From the patient's perspective, discrediting the patient or devaluing or failing to understand the patient's views only prevent interactions which give patients confidence that their treatment is successful, and over the longer-term, that they are cared about. Of course, there needs to be reciprocity in the warmth of the patient, but this is more likely to be engendered by an effective communication of concerns. In addition, medical interviews that are conducted with skill and a humanistic concern rather enhance the quality of the encounter and are more likely to be more efficient and effective diagnostically, but do not necessarily take more investment in time and energy than untrained interviews.

Figure 1. Showing factors influencing physician/patient communication.

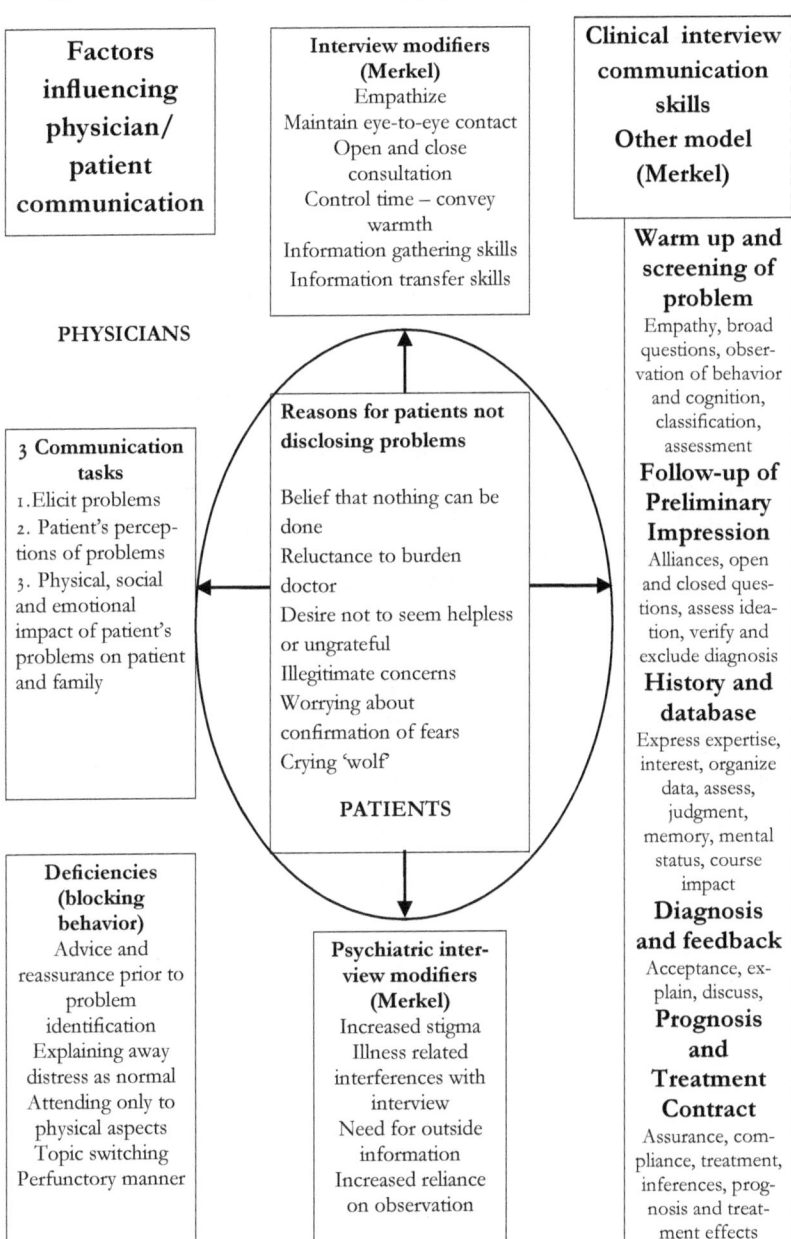

Factors influencing physician/ patient communication

PHYSICIANS

3 Communication tasks
1. Elicit problems
2. Patient's perceptions of problems
3. Physical, social and emotional impact of patient's problems on patient and family

Deficiencies (blocking behavior)
Advice and reassurance prior to problem identification
Explaining away distress as normal
Attending only to physical aspects
Topic switching
Perfunctory manner

Interview modifiers (Merkel)
Empathize
Maintain eye-to-eye contact
Open and close consultation
Control time – convey warmth
Information gathering skills
Information transfer skills

Reasons for patients not disclosing problems

Belief that nothing can be done
Reluctance to burden doctor
Desire not to seem helpless or ungrateful
Illegitimate concerns
Worrying about confirmation of fears
Crying 'wolf'

PATIENTS

Psychiatric interview modifiers (Merkel)
Increased stigma
Illness related interferences with interview
Need for outside information
Increased reliance on observation

Clinical interview communication skills Other model (Merkel)

Warm up and screening of problem
Empathy, broad questions, observation of behavior and cognition, classification, assessment
Follow-up of Preliminary Impression
Alliances, open and closed questions, assess ideation, verify and exclude diagnosis
History and database
Express expertise, interest, organize data, assess, judgment, memory, mental status, course impact
Diagnosis and feedback
Acceptance, explain, discuss,
Prognosis and Treatment Contract
Assurance, compliance, treatment, inferences, prognosis and treatment effects

The substance of medical interviews is to develop an understanding of nosologic and dynamic factors in the patient's pathology. Nosologic factors refer to figuring out what the patient is suffering from. Dynamic factors refer to the discovery of the biopsychosocial aspects of a patient's condition.

Acknowledgements: My thanks are due to Dr Lindy Boyd-Wilson for comments and suggestions

References

Foucault, M. (1988). *Madness and civilisation: A history of insanity in the Age of Reason*. (R. Howard, Trans.). New York: Vintage.

Kandel, E. R. (1998). 'A new intellectual framework for psychiatry'. *American Journal of Psychiatry, 155* (4), 457-469.

Kirmayer, L. J. (2006). 'Beyond the 'new Cross-cultural psychiatry: Cultural biology, discursive psychology and the ironies of globalization'. *Transcultural Psychiatry, 43* (1). Retrieved from: http://www.sagepub.com/cac6study/articles/Kirmayer.pdf

Lewis, S., Pea, R., Rosen, J. (2010). 'Beyond participation to co-creation of meaning: mobile social media in generative learning communities'. *Social Science Information.* 49 (3): 1–19. Retrieved from: http://www.stanford.edu/~roypea/RoyPDF%20folder/A169_Lewis-Pea-Rosen_SSI_2010.pdf

Perkins, J.J. & Sanson-Fisher, R. W. (1996). 'Increased focus on the teaching of interactional skills to medical practitioners'. *Advances in Health Sciences Education, 1*, 17-28.

Phillips, A. (2002). *Equals*. New York: Basic Books.

Phillips, A (2005). *Going sane. Maps of happiness*. New York: Harper Collins.

Waitzkin, H. (1984). 'Doctor-patient communication. Clinical implications of social scientific research'. *Journal of the American Medical Association, 252* (17), 2441-6.

Waymack, M. H. (2009). 'Yearning for certainty and the critique of medicine as "science".' *Theory of Medical Bioethics, 30*, 215-229.

6

COMMUNICATION CONTEXTS
The Courtroom Expert Witness:
Psychological and Legal Perspectives

This chapter explores the psychology of the expert witness in the court-room setting, gives a brief history of the psychological witness, the admissibility of witness statements, the contexts of witness statements in the court-room setting, the role of witnesses, and discusses problems with witness presentations.

The modern concept of the expert witness may have had its origins in Adam's Smith's notion of the 'Impartial Spectator' from his *The Theory of Moral Sentiments* (1759). Smith's notion had two elements in the pursuit of public reason: first, the desire not to omit relevant perspectives, and secondly, the desire to broaden discussion to avoid the localization of values. These are also both implicit aims of justice. The definition of an 'expert' is an individual with knowledge or experience on a subject, issue, or matter beyond that of a lay person. Dinur (2011) defines experts as " . . . those who the decision maker believes ha[s] the essential information, the distinctive competencies, and the proper decision framework needed in order to reduce the task uncertainty to an acceptable risk level in order to make a cogent decision" (p. 701). Thus an expert witness has what Sen (2010) calls an effective indirect power, exercising a remote form of control, a "capability without dependence" (p. 304). As another Smith (1989) suggests, the commonly accepted criterion for the testimony of an expert witness is whether it will "assist the trier of fact to understand the evidence or to determine a fact in issue" (p. 149).

In the legal setting, an expert is an individual who has knowledge of a subject, issue or matter beyond the knowledge of the jury. In New Zealand, Section 4 (1) of the *Evidence Act 2006* defines an expert as a "person who has specialized knowledge or skill based on training, study, or experience." Snook, Cullen, Bennell, Taylor and Gendreau (2012, p. 54) refer to the acceptance of information given by authority figures (or experts) as correct as the use of an 'expertise heuristic.' However, there may be notable exceptions, for example that of criminal profiling. Although Criminal Profilers have testified as expert witnesses in court, as Snook et al., (2012) explain that:

> Legal scholars, however, have been quick to challenge this notion because criminal profiling is not a generally accepted scientific technique, is not reliable, cannot prove the guilt of the defendant, and does not provide the explanations that are outside the normal understanding of the jury. (p. 54)

The roles of an 'expert' and a 'witness' are not synonymous. In psychological practice, in the court room there may be three categories of witness: a professional witness, an ordinary witness, or a witness of fact. The first and third roles are more common in court room practice involving psychologists. Gudjonsson (2003, p. 159) defines forensic psychology "as that branch of applied psychology which is concerned with the collection, examination, and presentation of evidence for judicial purposes." Mullen (2010, p. 169) further states that the, " . . . expert is caught between narratives of the law, politics, current social attitudes, the media, the psychiatric and behavioral sciences, and . . . the profession to which they belong."

Admissibility of Witness Statements

As Ireland (2008) states, expert witnesses are "expected to provide evidence on both facts and opinion, and to be clear about what type of fact they are basing their opinion on" (pp. 116-117). According to Blackwell (2011, p. 25), in New Zealand there are three main rules governing the admissibility of expert evidence. These are the expertise rule, the 'field of expertise rule,' and the basic rule – that an expert witness cannot provide evidence on the ultimate issue. All three rules are contained within Section 25 of the evidence Act 2006. Implicit here is that a statement of opinion is not admissible in a proceeding to prove the truth of what is believed – statements must be about facts.

With an overriding duty to the court, the legal brief for psychologists as witnesses does not involve conjecture. Psychologists giving expert witness statements in court need to be qualified in both content and process (Ireland, 2008, pp. 116-117). However as Roberts (2004) points out, in *Toohey versus Metropolitan Police*, a 1965 case in English law when the House of Lords held that an accused person should be permitted to adduce medical evidence as to the hysterical and unstable nature of the alleged victim of an assault in response to the admissibility of expert witness evidence, the presiding Justice Irving (as cited in Roberts, 2004, p. 229) stated:

> Because a citizen with something to offer as *amicus curiae* (someone not party to a case who volunteers to assist a court in its deliberations on a certain matter) comes to court by a particular route does not mean [that] he and his contribution are defined by the route he takes.

Justice Irving may be over-generous in his description of the courtroom witness' independence from the processes of the court, but he does emphasize the point that the courts call upon expert witnesses to shed light where none has been cast or to provide salience to the validity of a particular strain of argument.

While the expert witness must be conversant with court process, he or she is not necessarily bound by anything other than his or her own expert opinion and a responsibility to true and accurate science before the court. The expert witness needs to be able to exercise positional objectivity, the objectivity of what can be described from a specified position which may require 'interpersonal invariance' in fixed observational positions (Sen, 2010, p. 156). The case of *Toohey v Metropolitan Police* was also instrumental in differentiating between the kinds of evidence admissible. The judgment included the recognition that there is a distinction to be made between presenting evidence on expert psychological opinion that someone is incapable of giving reliable evidence and presenting psychiatric evidence that shows a witness is capable of giving reliable evidence but may not do so. This is a question for the jury to decide upon given the warnings of counsel and court (Roberts, 2004, p. 223). As Ryan (2003, p. 285) suggests, juries and judges are presumed to hold knowledge of "general fields of human endeavor and experience," and as such testimony on these matters by an expert witness is not relied upon.

It is not enough that the expert witness has an excellent knowledge in his or her field of study although this may be necessary but not sufficient criteria for being a legal witness. As Bond et al., (1999) claim, a professional witness also needs to be conversant in the court process, including the correct delivery of witness evidence. However, Gaughwin (2009, p. 156) claims that it should only be necessary to call on experts (for example, in mental health) where there is a disagreement on particular issues, in particular diagnosis and aetiology, and where prior agreement in pre-trial conference between experts cannot settle the dispute.

Consequently limitations on the role of the expert witness include not exercising expert opinion on matters of 'common knowledge,' not in areas outside their expertise, and not on the 'ultimate issue' before the court (Ryan, 2003, p. 285). In New Zealand, both the common knowledge and ultimate issue rule were abolished in the new Evidence Act of 2006 in favor of the 'substantial helpfulness' rule (Blackwell, 2011, p. 27). However as Komesaroff (2005) suggests, "researchers as expert witnesses may draw on their own and other's research to support their view and it is the role of legal counsel to explore, and in most cases, question those opinions" (p. 137). Concomitant with this is the necessity that "expert disagreement provides sensible containment boundaries" (McKenzie, van Winkelen & Grewal, 2011, p. 409).

A Brief History of the Psychological Witness

Ireland (2008, p. 117) claims that formal courts have used expert witnesses from the 1300s in Europe, mainly in the provision of interpreting medical terminology. Ryan (2003, p. 285) states similarly that the English court system has used expert statement in the medical context in the sixteenth century. This is a statement from an English judge in 1554:

> If matters arise in our laws which concern other sciences and faculties we commonly call for the aid of that science or faculty which it concerns, which is an honourable [sic] and commendable thing. For there by [sic] it appears that we do not despise all other sciences but our own, but we approve

them an encourage them . . . (Saunders, J. 1554, cited in Golan, 2004, p. 18)

Clearly the courts have called for testimony that is beyond the knowledge of judges and jurors since the foundation of jury trial. However, it is not until the nineteenth century in English law that psychological writing and pedagogy coincided with case law in defining the parameters of the psychological expert court witness.

The English case that decided court room rules concerning psychological competence was that of the case concerning the attempted assassination of the British prime Minister in 1843 by Daniel M'Naghten. M'Naghten fired a pistol, the bullet from which hit Robert Peel's secretary Edward Drummond in the back, killing him. The House of Lords considered a series of hypothetical questions relating to the trial involving insanity. Daniel M'Naghten was acquitted, although the trial laid the foundation for the M'Naghten Rules (despite the fact that had these rules been applied to M'Naghten's case he would not have been acquitted). The M'Naghten Rules have only gained status by common law but remain a standard test for criminal liability in relation to mentally disordered defendants. The rules state as follows:

> The jurors ought to be told in all cases that every man is presumed to be sane, and to possess a sufficient degree of reason to be responsible for his crimes, until the contrary be proved to their satisfaction; and that to establish a defence on the ground of insanity, it must be clearly proved that, at the time of the committing of the act, the party accused was labouring [sic] under such a defect of reason, from disease of the mind, as not to know the nature and quality of the act he was doing; or, if he did know it, that he did not know he was doing what was wrong. (1843, 10 Cl & Fin at 210)

The central question germane to the M'Naghten rules are whether the defendant knew what he or she was doing, and if he or she did, did they know that it was wrong? As such their application to specific circumstances sets out whether an accused may be judged not guilty by reason of insanity or guilty but insane. The sentence may involve a mandatory or discretionary period of treatment in a secure hospital instead of a prison sentence.

However, the psychological expert witness may not only give expert evidence on the issue of sanity but on any psychological matter pertinent to the case at the court's discretion. More recently, the Gillick competence test was added to the legal authority of decision-making for minors under 16 in the UK (see *Gillick v. West Norfolk and Wisbech Area Health Authority* 1986 A. C. 112). In terms of deciding whether to receive medical treatment, in order to be Gillick-competent, a young person needs to demonstrate "'sufficient maturity and intelligence' to understand the pros and cons of treatment" (Mackenzie & Watts, 2011, p. 49). The M'Naghten rules were followed in 1908 by Professor Hugo Munsterberg's publication *On the Witness Stand*, which further defined areas of concern for psychological expertise in legal settings (Ireland, 2008, p. 117). The Frye test was introduced following a 1923 judgment that allowed the admission of expert opinion only on the condition that it had gained general acceptance in the particular field in which it was related (Ryan, 2003, p. 286). However as Blackwell (2011, p. 26) states, "that test has now been rejected both in USA and New Zealand, in favour [sic] of a more expansive approach assessing the relevance and reliability of proposed evidence."

As Klee and Friedman (2001) relate:

> Just when a scientific principle of discovery crosses the line between the experimental and demonstrable stages is difficult to define. Somewhere in this twilight zone, the evidential force of the principle must be recognized, and while the court will go a long way in admitting expert testimony deduced from a well-recognized scientific principle or discovery, the thing from which the deduction is made must be sufficiently established to have gained general acceptance in the particular field in which it belongs. (p. 79)

That qualified psychologists could give testimony in court as experts on matters of 'state of mind' received further support from the US Court of Appeals for the District of Columbia Circuit decision in Jenkins v. United States (1962) (Bartol & Bartol, 2012, p. 3). Throughout the mid to late twentieth century, the use of psychological expert witnesses in court settings increased such by the 1980s that they were commonplace within the American court system. The case of *Barefoot v Estelle* (1983) established a US supreme court ruling that expert psychiatric opinion could be admitted in court despite an

Amicus Curiae brief by the American Psychiatric Association in support of the defendant's position that such evidence was inadmissible. This has led other commentators such as Smith (1989, p. 145) to postulate that the courts rely on a broad range of informed speculation, "which is too often more speculation than informed." As Roberts (2004, p. 215) relates, the case of *R v MacKenney and Pinfold* (1981) is considered a well-known authority on the admissibility of expert witness evidence in the adversarial system because it also contains a condition under which a judge refused to accept an 'expert' challenge to a prosecution witness' credibility. The 1990s brought two further landmark court decisions on the provision of psychological evidence in court: Lipsker's case (1992) and *Daubert vs. Merrell Dow Pharmaceuticals* (1993). Lipsker's case determined that 'recovered memory' evidence under hypnosis prior to the giving of oral evidence was not admissible in court, whereas Daubert's case led to the development of the Daubert Criteria.

The 'Daubert criteria' established the importance of psychological experts being explicit about the methodologies and measures they employ and the scientific foundations of their opinions, regardless of the issue and context they are presenting evidence within (Ireland, 2008, p. 118). The Daubert criteria aim to separate scientific fact from opinion or conjecture. The four criteria involve the concepts that if evidence is scientific, it needs to be: falsifiable (a product of testable technique), reviewed (subject to peer review in professional journals), accepted (the theory or technique must have general acceptance in the community), and have a known error rate (Ireland, 2008, p. 118). A psychological witness submits evidence on scientific principle and not merely outcome of specialized opinion. This scientific principle may be seen as a positive motivation for the accuracy of the expert witness opinion, however, so called 'negative motivations' also exist. For example, in English law the offence of perjury consists of the prescribed conduct of making a statement under oath in a juridical proceeding, with the *mens rea* (or mental culpability) of knowing this statement to be false (Ormerod, 2008, p. 45). Consequently there is considerable investment in the expert witness as arbiter of truth in the legal setting. However, in New Zealand, as Blackwell (2011) suggests, "It will be for judges to continue to be the gatekeepers in terms of admissibility of expert evidence in the context of criminal and civil proceedings" (p. 26).

Contexts of Witness Statements in the Courtroom Setting

There are two systems of legal procedure: the adversarial and the inquisitorial system. The occasional medical expert witness needs to have knowledge of the adversarial court system and the role of the medical expert within it (Ryan, 2003, p. 283). The adversarial system relies on the presentation of evidence in a structured and stylized format, characterized by evidence in chief and cross-examination. In New Zealand, a re-examination may follow "if the prosecutor wishes to follow up evidence that has from cross examination" (New Zealand Psychology Society, 2010, p. 2). The system is devised to exclude unreliable evidence. The psychological expert qualifies as expert by virtue of the fact that he or she has a specialized knowledge not held by the general public. His or her role is sometimes to offer an opinion as evidence. There are limitations to the role however; Ryan (2003, p. 283) states that they are the limitations of common knowledge, field of expertise, and the ultimate issue.

Consequently, there are different fields of access to information in the medical setting and in the adversarial legal setting. The exchange of information between and among medical health professionals is bound by a different code of ethics than that among lawyers, although principles of truth, verifiability, and veracity may be similar but not the same. The Turner case (1975) determined the limitations with which psychologists and psychiatrists could be called on by the courts to prove the probability of the veracity of the accused. If not for the Turner case, medical experts could then arguably take the place of trial by jury (Roberts, 2004, p. 215). Here is a comment from Lawton's judgment in that case (as cited in Roberts, 2004, p. 215): "[W]e are firmly of the opinion that psychiatry has not yet become a satisfactory substitute for the common sense of juries or magistrates on matters within their experience of life."

A further issue for the adversarial court process may be that of cultural differences in the language interpretation/translation context. For example Lee (2009) has argued that the "absence of cultural and/or linguistic explanations by the interpreter, evidence given by witnesses from culturally and linguistically diverse backgrounds may not be accurately or fully interpreted, and this can have potentially serious consequences for the witness in the adversarial context" (p. 379). Although the notion of the linguistic 'conduit' has attempted to dispel the extent to which the interpreter or translator influences interpretation of testimony (see Gaio v R (1960) 104 CL 419 at 429 Fullagar J) cultural and linguistic differences remain an ongoing and problematic issue for the court process. Interpreters and translators are now seen more as facilitators of communication in a courtroom setting (Lee, 2009, p. 380).

Medics, psychologists, and physicians require relatively free communication as the basis of clinical problem solving; this exchange takes place between both patients and colleagues and is a part of everyday professional life. As Ryan (2003) suggests, "In the curial interface between medicine and law, science is moulded [sic] into an uneasy and unnatural use within a system of intellectual combat alien to [the] scientific method. Enmeshed in this process is the expert witness" (p. 283). The court procedure of the adversarial system is stylistically defined by an examination on behalf of the party by whom they were summoned, termed 'evidence in chief.' Subsequently, the expert psychological witness will be questioned by the opposing party under cross examination. The evidence is tested by adversarial probing for accuracy, consistency, or veracity.

Despite the fact that from a sociological perspective expert opinion brings into the courts a "discourse not about the offence itself . . . but the offender's psychological, moral and social functioning" (Mullen, 2010, p. 166), the expectations of the expert witness are that they provide black and white, unambiguous responses to questions about which there is increased uncertainty (Ryan, 2003, p. 284). On the one hand there may be pressure to simplify and consequently distort information; on the other hand elaborate terminologies may prevent equivocal understandings. Furthermore the adversarial system of courtroom procedure may be different in process from the witnesses' normal professional lives, leading to compromised expression or opinion. However, as Ryan (2003, p. 284) states, the medical witness in the adversarial court system may be frequently concerned with probabilities rather than with absolute truth, i.e., proof 'beyond reasonable doubt' of legal liability. The inquisitorial system may be more beneficial in respect to medical cases and family law, from the perspective that it is not as divisive.

Further criticisms of the adversarial court system in terms of expert witness are that it may encourage partisan and hence biased testimony, facilitate paid experts under financial incentive, or lead to over-simplification, confabulation, or even deliberate omission (Gaughwin, 2009, p. 159). Consequently, the status of the evidence of the expert witness under the adversarial systems runs the risk of being considered inherently flawed, underscoring the need for the Daubert Criteria. Gaughwin (2009, p. 157) puts it more simply:

In the forensic context, a report will only be as good as the material the expert has to work with. This is particularly so in those situations where the expert is asked to review another expert's report, but does not examine the person in question.

Klee and Friedman (2001, p. 80) explained that the judge may act as gatekeeper in determining rules of evidence based not only on general acceptance but also determinations of quality and technique. Consequently they claim that the court has recognized that science is a process just as the legal system is a process, relying on agreed norms of behavior and systems of investigation.

Roles of Witnesses

Despite the fact that as Foucault (2003) has claimed that the medical expert's legitimacy and power is derived from the authority of science rather than judicial power (whose legitimacy is derived from the state), the role of the expert psychological witness is informed by two professional contexts: that of the courts and that of the psychologists' own professional body. The position is seen as being one of privilege defined by the courtroom setting, not necessarily in terms of a special honor other than owing a responsibility to the truth to two contexts of authority.

According to Swann (2002, pp. 305-308), there are eight main duties of the expert witness in regard to the evidence he or she provides:

Expert evidence is independent	Expert evidence presented to the court should be seen to be the independent product of the expert, uninfluenced by the circumstance of litigation
Expert evidence is objective and unbiased	An expert witness provides independent assistance to the court in the form of objective unbiased opinion in relation to his or her own expertise
Facts behind opinions must be asserted	An expert witness is under duty to assert facts or assumptions upon which his or her opinion is based. Material facts should not be occluded.

The extent of expertise must be explained	An expert witness should make it clear when questions or issues fall outside his or her expertise and indicate when an opinion is provisional.
Limitations of opinions must be asserted	If an expert's opinion lacks the necessary research because of insufficiency of data, then the provisional nature of the opinion should be asserted. Experts should not mislead by omissions.
Change of opinions must be communicated	If, consequent to the exchange of reports, an expert witness changes his or her view, this change of opinion should be communicated without delay to the courts.
Supporting documentation must be provided to all parties	If the expert evidence is reliant upon supplementary documentation such as photographs, plans, calculations, or survey reports, these must be provided to the opposite party at the same time in the adversarial court room context.
Departures from normal conduct must be reported	If, contrary to normal practice, an expert does give an opinion that promotes a particular cause or provide a report that is not wholly objective, the report must make these limitations clear.

Swann (2008, p. 309) describes the judge's possible view of expert psychological witnesses as being constrained by three main factors: the possibility of misleading the court, the possibility of the expert witness as being unable to express a firm opinion on a particular happenstance relevant to the case, and the expert witness possibly being unable to express an opinion on whether or not someone is telling the truth.

Smith (1989) notes that there are significant grey areas in the acceptance and or admissibility of expert evidence by the Anglo/American courts. One significant concern is the reluctance in common law to permit direct or unchallenged testimony on the truthfulness or otherwise of another witness; authorities can be found for and against this proposition. The courts are also divided over the issue of the proffering of expert opinion in issues of syndromes and profiles, particularly in circumstances where the psychology of their understanding is new or novel in approach. There are generally understood to be five operative reasons for the court's reluctance to accept evidence from an expert witness, which operate at a level of abstraction behind the Daubert Criteria (Smith, 1989, p. 161). These are questions concerning the reliability of evidence, such as:

1. What are the necessary tests for evidence?

2. Does the evidence fall within the common experience of the jurors? (Can they understand it in context of the case?)

3. Is the evidence misleading for the jury or does it invade the province of the jury? (Does it allow the jury to perform the necessary truth-finding task?)

4. Are there improper comments on the credibility of the witness?

5. Is the novelty of testimony strained by the possibility for 'popular' or 'junk' science to be applied regarding a legal question or issue?

Hence, problems the courts face center on the possibility of incorrect decisions. Incorrect testimony from an expert witness arguably does more harm than inexpert lay-testimony because of the scientific status pertaining to the expert within the court system. Further issues concern those of the burden of proof or the possibility for the establishment of legal rules on faulty evidence or scientific inferences (Smith, 1989, p. 162). Likewise, testimony that is inefficient – irrelevant, repetitive, or incomprehensible – is likely to be viewed as interfering in the operation of the courts. On the other hand, the possibility for cross-examination in the adversarial court system has arguably resulted in the admissibility of psychological expert evidence of low accuracy. Furthermore, the admissibility of expert witness opinion may also result in the introduction of hearsay or in the introduction of biased accounts motivated by payment for a particular cause. Finally, psychological expert witnesses are unable to comment on the 'ultimate issues' of the case, as these are the provenance of the judge and jury (Smith, 1989, p. 166).

Styles of Questioning

Roth and Mehta (2002) argue that there are generally believed to be three main human influences on reported data quality.

Table 1. Influences on human error.

Memory	Forgetting details
	Memory alterations
Problems of vested interests	Protection of professional status or personal reputation
	Political influence on outcome
	Protection from blame
	Psychological interest in maintaining self-esteem
Mistaken judgments	Honest mistakes
	Faulty inferences based on incomplete information
	Acting on second- or third-hand information stemming from erroneous sources
	Obfuscation – blurring details
	Incorrect inferences from faulty heuristics

Needless to say, the function of the expert witness is to report the truth and to dispel any of the above. However, Wheatcroft and Woods concluded in a paper published in 2010 that there has been little research examining the effects of witness preparation for accuracy in court reporting (p. 187). However, witness-training programs do exist which aim to identify and prepare expert witnesses to identify tactics and questioning styles used by lawyers in cross-examination. This is premised by the fact that in the adversarial system the rationale of cross-examination may be to discredit the opposing testimony (Wheatcroft & Woods, 2010, p. 187). The three main elements of this preparation involve listening carefully to questions, requesting clarification, and not answering a question that is not understood. Wheatcroft and Woods (2010, p. 190) suggest that there are two main findings of witness preparation: firstly, warning participants about misleading information may result in lower compliance to suggestive effect, and secondly, witness accuracy may be improved by recognition that the courtroom context may produce possible attempts to mislead. Cross examination is used to probe the accuracy of witness accounts or evidence obtained in examination-in-chief, in order to expose inaccurate or unreliable witness statements. It is a trend that witness reliability is shown to drop during cross examination. Furthermore, if lawyers question witness testimony, juries may be less likely to find the defendant guilty (Wheatcroft & Woods, 2010, pp. 191-192). The one element of eyewitness testimony which is immune from external examination is that of memory, but even this is not absolute.

A second major factor in preparing witnesses for court testimony is familiarity with the concept of the leading question. These are censured in examination-in-chief but permitted in cross-examination. The main feature of leading questions is that they are suggestive in nature and aim to manipulate answers by limiting responses to a two-choice alternative, or to elicit preferred answers (Wheatcroft & Woods, 2010, pp. 191-192). Similarly, a firm rather than friendly approach to re-questioning is more likely to result in the alternation of initial responses. However, this understanding reinforces the need for witness preparation in courtroom procedure as questions are subject to very little overall regulation (Wheatcroft & Woods, 2010, p. 194). Confidence and accuracy are seen as the two most desirable traits of the expert witness as this underscores witness credibility for the jurors; however, the two are not always necessarily closely correlated. Putting witnesses on-guard through preparation means that they are less likely to be misled during re-questioning. However, the fact remains that leading questions may be likely to reduce witness accuracy (Wheatcroft & Woods, 2010, p. 203). Techniques used for discrediting an expert witness under cross-examination include pointing out weaknesses in a line of argument, highlighting contradictions between the expert's views and other sources, and undermining the witness's expertise.

For court witnesses with disabilities, the type of questioning that is asked also has a significant impact on testimony. Specific leading or complex questions can cause problems of confabulation, suggestibility, and acquiescence (Kebbell, Hatton, Johnson & O'Kelly, 2001, p. 99). Confabulation involves replacing gaps in memory with distorted material. Suggestibility involves providing an overly-compliant answer to a question that is believed to be required by the questioner. Acquiescence concerns the fact that people may be more likely to respond either yes or no to directive questions, limiting the response (Kebbell et al., 2001, p. 99). Questions which communicate the answers required are directive questions, while confusing questions may be composed of negative or double negative suggestions, multiple questions, or complex questions in terms of sentence structure (Kebbell et al. 2001, p. 99). Directive questions are the most common type, followed by questions involving negatives (Kebbell et al., 2001, p. 99).

Problems with Witness Presentations

Freckelton (2007, p. 185) points out that the witness immunity rule provides protection from a variety of civil actions against expert witnesses, including for those who write reports, undertake preliminary work, and give evidence. There is an ongoing issue of whether this rule should be extended to include protection against sanction by psychologists' own disciplinary bodies for those who give court evidence. As Freckelton (2007) states: "The long-standing and often repeated stance articulated by the common law is that no action will lie against a witness for words spoken in the course of giving evidence" (p. 186). The immunity from civil actions is based on two main principles: 1.) a 'free and fearless' giving of evidence, 2.) to avoid further contestation of the truth of the evidence (Freckleton, 2007, p. 187). Thus, the confines of the witness box prevent attempts at re-litigation. Freckleton (2007, p. 193) points to fifteen possible ways in which expert witness evidence may be compromised and provides the case in which these limitations were established:

A failure to adhere to the obligations of neutrality, non-partisanship and even-handedness in the forensic role.	Mustac v Medical Board [2004] WASCA 156 This case involved an appeal against improper conduct with respect to the methodology in the writing of two forensic reports. A psychometric test was applied to questions of veracity when it was not designed for that purpose.

Misusing tests to achieve a particular result.	Mustac v Medical board [2004] WASCA 156
Adoption of advocacy role	General Medical Council v Meadow [2006] EWCA Civ 1390; [2007] 1 All ER 1; This case was an appeal against ruling that a) an expert witness had not been guilty of serious professional misconduct, and b) that professional expert witnesses should be offered some immunity from prosecution by their regulatory bodies for evidence presented in good faith in court. Appeal in a) was dismissed but in b) allowed. Council for Regulation of Healthcare Professionals v General Medical Council [2005} EWHC 579 (Admin). The issue of undue leniency by a professional medical body was examined.
Exceeding the parameters of expertise	General Medical Council v Meadow [2006] EWCA Civ 1390
Utilizing unscientific methodologies	General Medical Council v Meadow [2006] EWCA Civ 1390
Adoption of unwarrantedly inflammatory language:	General Medical Council v Meadow [2006] EWCA Civ 1390; [2007] 1 All ER 1
Dishonesty	Hussein v William Hill Group [2004] EWHC 208 Explored the issue of witness immunity and conditions under which expert evidence is unacceptable due to undeclared interest or association.

Detracting from the reputation of the profession	Re Watson-Munro [2000] PRBD (Vic) 4 The issue of illegal incentive for testimony was explored. (A substance dependent forensic psychologist provided cocaine for a lawyer who gave him forensic work – in relation to substance-dependent offenders).
Breaching client confidentiality without justification	Re Noble [2002] PRBD (Vic) 6 The liability and veracity of expert witness reporting was examined. A psychologist wrote a report and gave evidence in a family court of dangerousness to custodial children of fathers who are homosexually active. "Draft report written for him, research not undertaken by him, views contrary to state of knowledge" – Freckleton (2007)
Grossly incompetent analysis, diagnosis or prognosis	Re Noble [2002] PRDB (Vic) 6
Misleading a court or tribunal	Re Noble [2002] PRDB (Vic) 6
Failing to provide a court with accurate and documentable information	Austin v American Association of Neurological Surgeons, 253 F 3d 967 (7th Circuit, 2001). The issue of deliberate unprofessional censure in relation to professional medical membership was examined. The issue in this case was whether a medical society could discipline one of its members for testifying falsely as an expert witness. Censorship was upheld.
Engaging in unjustified conjecture	Council for regulation of Healthcare Professionals v General Medical Council [2005] EWHC 579 (admin) A claim of undue leniency by a medical authority in response to malpractice was upheld.

Assumption of minority or heterodox views, without disclosing their status:	Re Paterson, unreported, professional conduct Committee, General medical Council, 3 April 2004: http:www.gmc-uk.org/probdocs/decisions.pcc/2004/PATERSON_20040304.htm The issue that it is the expert's responsibility to ensure that an expert's report is properly researched and substantiates any departures from existing published work was explored.
Departing from their own published views without justification and explication	Re Paterson, unreported, Professional Conduct Committee. General Medical council, 2 April 2004: http:www.gmc_uk.org/probdocs/decisions.pcc/2004/PATERSON/_20040304.htm

For lawyers protecting the interest of their client, or prosecutors presenting a case against the defendant, expert witness credibility is undermined by three main factors: the expert unexpectedly reneging on an opinion, an expert showing biases or errors in judgment, and an 'expert' with inadequate qualifications (Leslie, Young, Valentine & Gudjonsson, 2007, p. 401). Conversely, clarity of language, firm conclusions, and a clear presentation and demeanor are identified by Leslie et al., (2007, p. 400) as qualities which make an expert good at giving court-room evidence. However, the position of the law regarding witness immunity may be changing. In the recent case of Jones v Kaney [2011] UKSC 13 the United Kingdom Supreme Court "abolished the principle of expert witness immunity from suit in relation to evidence in a 5 to 2 majority decision" (Green 2011, p. 1). Less recent cases in New Zealand have applied witness immunity to negligence claims against expert witnesses (B v Attorney General [1997] NZFLR 550 (FC) and RIG v Chief Executive of the Ministry of Social Development HC Auckland CIV-2008-404-0033461). However, in Lai v Chamberlains [2007] 2 NZLR 7 the court abolished immunity from liability in negligence for advocates in New Zealand (Green, 2011, p. 3).

Restorative Justice and the Expert Witness

According to Woolford and Ratner (2010, p. 6), "restorative justice is the umbrella term for programs that seek to involve victims, offenders, and community members in addressing the harms caused by crime." However in the New Zealand/Aotearoa context, as Cooper, Rickard and Waitoki (2011, p. 37) suggest, "there is no legislated definition of what restorative justice is, or how meetings (conferences) should be executed." Rather, according to the Ministry of Justice (2010, p. 1), restorative justice is "based on a set of principles in process." Restorative justice differs from retributive justice because the former "essentially refers to the repair of justice through unilateral imposition of punishment, whereas restorative justice means the repair of justice through reaffirming a shared value-consensus in a bilateral process" (Wenzel, Okimoto, Feather & Platow, 2008, p. 375). It is seen as an alternative to the punitive and professionalized contexts of the retributive justice system and is defined by values of empowerment, healing, and openness and by practices such as face-to-face interaction and open dialogue.

Crucial to this process is healing rather than punishing – healing the victim, the moral and social self of the offender, and the community (Wenzel et al., 2008, p. 376). Here, the Indian concept of justice as *nyaya*, (which stands for a comprehensive concept of realized justice) is useful. As Sen (2010, p. 20) points out, *nyaya*: ". . . is inescapably linked to the world that actually emerges, not just the institutions or rules we happen to have." This is much closer to the 'redistributive' outcome-focused nature of restorative justice. In the New Zealand context, restorative justice must be about a form of power sharing, at least from an inter-cultural perspective. The redistribution of power and its experience, at least symbolically, is thus potentially a process that restorative justice shares with retributive justice. As Cooper, Rickard and Waitoki (2011) explain,

> Attaining as level of cultural competency and cultural safety in New Zealand requires acknowledging that a power imbalance exists in society that is detrimental to Maori when it is unevenly distributed. Power-sharing relates to Maori being allowed to share knowledge, language, space, ideas, practices, worldviews, and pedagogy." (p. 51)

A founding principle of the restorative justice vision is that of Christie's view (as cited in Wenzel et al., 2008, p. 377) that offenses are considered conflicts that "rightfully belong to victims and offenders." There are two values which transcend both retributive and restorative justice processes, behavior control and justice restoration (Wenzel at al., 2008, p. 378). Retributive justice is more prevalent when the victim and offender lack a common identity, and restorative justice when a common identity is shared (Wenzel et al., 2008, p. 383).

However, just as with the retributive justice system, the limiting character of the informal-formal justice complex is becoming increasingly under the sway of neoliberalism and juridification. Despite this, one might contend that restorative justice is closer in conception to the Rawlsian (1971) concept of the justice of fairness derived from an understanding of an original position. This is the imagined situation of equality or the 'veil of ignorance,' when all is equinanimous and parties are free to choose the 'good life.' The application of Rawls's concept is toward defining just institutions. If we invert the Rawlsian concept to claim that this is seen as the goal of restorative justice in which parties may amicably go their separate ways, Rawls's description of the means to the end needs to also be inverted. Contrary to the sustained dialogue necessary for restorative process, the Rawlsian concept requires that the parties have no knowledge of their personal identities, or their interests within the group prior to making their decisions about fairness – which is closer to the perspective of the expert witness than to the process of justice itself in restorative forms.

The difference is that Rawlsian objectivity retains a remnant of utilitarianism in its intent to describe an institution of justice (despite the connotations of metaphorical borrowing from the concept of marriage). Further, while the process of restorative justice is institutional it is also inherently deontological. Rawls's approach does agree with the restorative justice approach in that, as Sen (2010, p. 69) observes, it attempts to combine "the operation of the principles of justice with the actual behavior of people." Arguably it differs on four counts: first, it ignores comparative questions of justice by focusing on a perfectly just society; second, by concentrating on 'just institutions' it ignores other forms of social realizations in the name of justice; thirdly, it may fail to account for the institutional necessity of culturally diverse voices; and fourth, it fails to take an institutional procedural account of parochial values (Sen, 2010, p. 90). Accommodation of these four factors are additional ways that restorative justice differs from retributive justice.

The nexus of restorative justice origins is the need to understand how language applies to concepts of value in a political economy and how these might flow from legal concepts of human rights. Just as with the retributive justice system, the Foucaultian (2003, p. 8) perspective regards the conflict resolution setting as characterized by 'techniques of discipline' (in so much as the process needs to reflect the grievance, the reparative dialogue, and the measure of justice exchanged) and 'technologies of the self'. As Foucault (1988) suggests, technologies of the self:

> Permit individuals to effect by their own means or with the help of others a certain number of operations on their own bodies and souls, thoughts, conduct, and way of being, so as to transform themselves in order to attain a certain state of happiness, purity, wisdom, perfection, or immortality. (p. 18)

The latter of course may partially be the goal of justice, certainly of the restorative variety for the victim if not always for the offender. Similarly, Habermas' (2010) discourse on the public sphere and the ideal speech situation (in which discourse on public issues is entered into by people who meet as equals, regardless of their social status or nationality) defines the context of restorative justice as infused with a power matrix that nevertheless creates conditions for "non-coercive and self-corrective deliberations" (Cited in Woolford & Ratner, 2010, pp. 8-9). The role of the courtroom witness is itself to some extent that of providing the court with evidence from the position of the 'ideal speech situation' through independence from the perspective of both offender and victim.

Woolford and Ratner (2010, p. 10) also define three positions in which restorative justice attempts to remove underlying causes of oppression and injustice: The first of these is 'informal justice retreatism' – which seeks to provide a nomadic space removed from the informal-formal justice complex. The expert witness to some extent occupies this space within the regulation of the court-room. The second strategy is the 'trickle-up' model, in which the intention is to achieve justice transformation from within the informal-formal justice complex. The expert-witness clearly occupies an open position with regard to this internal process. The third position is called the 'informal justice counterpublics' – a communicative space out of which it may be possible to challenge the informal-formal justice complex. The court-room expert witness may also play a role in this process, depending on the context and nature of the opinion given to the court, even if it is only a character testimony.

The expert witness is partially about the creation of exceptions in the courtroom setting or exceptions to the regulatory framework of the court. This fits quite neatly with the expectations that in restorative justice, transgressions (about which dialogue is entered into) are a violation of values shared between victim and offender because of membership of a certain community. The impartial testimony of an expert witness may also ameliorate the alienation effect of the courtroom setting and be neutral with regard to symbolic compensation (Dazur, 2003, p. 284). Finally, the expert witness is subsumed within but not dictated by the legal substance of the court. If advocates of rehabilitative justice reject professional control, then the expert witness may be neutral with regard to this power dialectic of, for example, family group conferences, victim-offender reconciliation programs, or reparative boards, and multicultural practices and could potentially take on many shapes and forms (Dazur, 2003, p. 279). The relation of Maori culture to restorative justice is a relatively under-explored phenomenon – at least in academic texts. However, as Cooper, Rickard and Waitoki (2011) suggest, " . . . there is increasing interest from Maori in the commonalities between the philosophical approaches of therapeutic jurisprudence and tikanga Maori, and the potential for improving outcomes for Maori through the combined use of these in the justice system" (p. 38). Tikanga is guided by the principle of respect for the dignity of person and peoples, hence it is inherently valuable in considerations of restorative justice.

Conclusion

The psychology of the courtroom witness has evolved in parallel with the court system in Anglo/American jurisdictions. Historical record shows that expert witnesses were used in court room setting since the origin of the jury trial in the Europe of the middle ages. From the late nineteenth and mid-twentieth centuries in particular, the standards, criteria, expectations, and rules under which the expert psychological witness may give testimony have become simultaneously more exacting as the range of testimonial evidence topics has become broader. Psychologists may need to be trained in the styles and procedures of giving courtroom evidence and expert opinion, according to the protocol of the jurisdictions under which they serve. It is clear that psychologists giving expert witness evidence have a duty both to the court and to the professional body which accredits them, although while giving evidence their duty to the court will override that of the professional body to which they belong. Concomitant with this duty is the need to satisfy the Daubert criteria in giving evidence. However, Mullen (2010) cautions that if we view medical science as just another discourse of power while at the same time as idealizing the capacity of the expert witness to be an absolute arbiter of truth, then we may also become fools "when we leave science behind and join the 'morality play'" (p. 175) of the courtroom. Finally, there may be some truth as the physicist Niles Bohr suggests (as cited in Lehrer, 2009, p. 55), that an expert is "a person who has made all the mistakes that can be made in a very narrow field."

References

Bartol, C. R., & Bartol, A. M. (2012). *Current perspectives in forensic psychology and criminal behavior.* Thousand Oaks California: Sage Publishing.

Blackwell, S. (2011). Expert evidence: The conduct of expert witnesses. In Seymour, F., Blackwell, S., Thorburn, J. (Eds.). In *Psychology and the Law in Aotearoa New Zealand.* Wellington, New Zealand: The New Zealand Psychological Society.

Bond, C., Solon, M., Born, S., Harper, P. (1999). *The Expert Witness in Court: A Practical Guide.* Crayford: Shaw.

Christie, N. (1977). 'Conflicts as property'. *British Journal of Criminology, 17,* 1–15.

Cooper, E, Rickard, S., Waitoki, W. (2011). Maori, psychology and the law: Considerations for Bicultural Practice. In Seymour, F., Blackwell, S., Thorburn, J. (Eds.). *Psychology and the Law in Aotearoa New Zealand.* Wellington, New Zealand: The New Zealand Psychological Society.

Dinur, A. R. (2011). 'Common and un-common sense in managerial decision making under task uncertainty'. *Management Decision, 49* (5), 694-709.

Dazur, A. W. (2003). 'Civic implications of restorative justice theory: Citizen participation and criminal justice policy'. *Policy Sciences, 36,* 279-306.

Evidence Act 2006 No 69 (as at 29 June 2009), Public Act. Retrieved from: http://www.legislation.govt.nz/act/public/2006/0069/13.0/DLM393463.html

Foucault, M. (1988). Technologies of the Self. In Martin, L.H. et al (1988) *Technologies of the Self: A Seminar with Michel Foucault*. London: Tavistock.

Foucault, M. (2003). *Abnormal: Lectures at the College de France 1974-1975* (Burchell, G., Trans.). New York: Picador.

Freckelton, I. (2007). 'Expert Witness Immunity and regulation of experts'. General Medical Council v Meadow [2006] EWCA Civ 1390; [2007] 1 All ER 1. *Psychiatry, Psychology and Law, 14* (1), 185-193.

Golan, T. (2004). *Laws of Men and Laws of Nature. The History of Scientific Expert Testimony in England and America*. Cambridge, MA: Harvard University Press.

Green, J. (2011). Is expert immunity in New Zealand on the brink?. Retrieved from: http://www.buildingdisputestribunal.co.nz/site/buildingdisputes/files/BuildLaw/Issue%2010/Is%20expert%2 owitness%20immunity%20in%20NZ%20on%20the%20brink.pdf

Ireland, J. L. (2008). 'Psychologists as witnesses: Background and good practice in the delivery of evidence'. *Educational Psychology in Practice, 24* (2), 115-127.

Gaughwin, P. C. (2009). 'Choosing and instructing the mental health witness: A lesser known aspect of tevorrow'. *Psychiatry, Psychology and Law, 16* (1), 155-162.

Gudjonsson, G. H. (2003). 'Psychology brings justice: The science of forensic psychology'. *Criminal Behavior and Mental Health, 13*, 159-167.

Kebbell, M. R., Hatton, C., Johnson, S. D., O'Kelly, C. M. E. (2001). 'People with learning disabilities as witnesses in court: What questions should lawyers ask?' *British Journal of Learning Disabilities, 29*, 98-102.

Klee, C. H., & Friedman, H. J. (2001). 'Neurolitigation: A perspective on the elements of expert testimony for extending the Daubert challenge'. *NeuroRehabilitation, 16*, 79-85.

Komesaroff, L. (2005). 'Going to court over education: Researcher as expert witness'. *Education and the Law, 17* (4), 137-153.

Lee, J. (2009). 'When linguistic and cultural differences are not disclosed in court interpreting'. *Multilingua, 28*, 379-401.

Lehrer, J. (2009). *The decisive moment: How the brain makes up its mind*. Melbourne: Griffin Press.

Leslie, O., Young, S., Valentine, T., Gudjonsson, G. (2007). 'Criminal barristers' opinions and perceptions of mental health expert witnesses'. *The Journal of Forensic Psychiatry and Psychology, 18* (3), 394-410.

Mackenzie, R., & Watts, J. (2011). 'Can clinicians and carers make valid decisions about others' decision-making capacities unless tests of decision-making competence and capacity include emotionality and neurodiversity?' *Tizard Learning Disability Review, 16* (3), 43-51.

McKenzie, J., van Winkelen, C., Grewal, S. (2011). 'Developing organizational decision-making capability: A knowledge manager's guide'. *Journal of Knowledge Management, 15* (3), 403-421.

Ministry of Justice. (2010). *Restorative justice in New Zealand: A summary paper.* Wellington: Ministry of Justice.

Mullen, P. E. (2010). 'The psychiatric expert witness in the criminal justice system'. *Criminal Behavior and Mental Health, 20,* 165-176.

New Zealand Psychological Society. (2010). 'Psychologists as expert witnesses: Guidelines concerning modes of evidence applications'. Retrieved from: http://www.psychology.org.nz/cms_show_download.php?id=949

Ormerod, D. (2008). *Smith and Hogan Criminal Law: Cases and Materials.* New York: Oxford University Press.

Rawls, J. (1971). *A Theory of Justice.* Cambridge, M.A: Harvard University Press.

Roberts, P. (2004). 'Towards the principled reception of expert evidence of witness credibility in criminal trials'. *The International Journal of Evidence and Proof, 8,* 215-232.

Roth, W. D., & Mehta, J. D. (2002). 'The Rashomon Effect: Combining positivist and interpretivist approaches in the analysis of contested events'. *Sociological Methods and Research, 31* (2), 131–37.

Ryan, M. (2003). 'The adversarial court system and the expert medical witness: The truth whole truth and nothing but the truth?' *Emergency Medicine, 15,* 283-288.

Sen, A. (2010). *The Idea of Justice.* London: Penguin.

Smith, A. (1790/1976). *The Theory of Moral Sentiments.* Oxford: Clarendon Press.

Smith, S. R. (1989). 'Mental health expert witnesses: Of science and crystal balls'. *Behavioral Sciences & the Law, 7* (2), 145-180.

Snook, B., Cullen, R. M., Bennell, C, Taylor, P. J., Gendreau, P. (2012). The criminal profiling illusion: What's behind the smoke and mirrors? In Bartol, C. R., & Bartol, A. M. (Eds.). *Current Per-*

spectives in Forensic Psychology and Criminal Behavior. Thousand Oaks California: Sage Publishing.

Swann, A. (2002). 'The roles and duties of the expert witness'. *Child Care in Practice, 8* (4), 305-311.

Wenzel, M., Okimoto, T. G., Feather, N. T., Platow, M. J. (2008). 'Retributive and restorative justice'. *Law and Human Behavior, 32,*. 375-389.

Wheatcroft, J. M., & Woods, S. (2010). 'Effectiveness of witness preparation and cross-examination non-directive and directive leading questions styles on witness accuracy and confidence'. *The International Journal of Evidence and Proof, 14,* 187-207.

Woolford, A., & Ratner, R. S. (2010). 'Disrupting the informal – formal justice complex: On the transformative potential of civil mediation, restorative justice and reparations politics'. *Contemporary Justice Review, 13* (1), 5-17.

Cases:

U.K

Council for Regulation of Healthcare Professionals v General medical Council [2005] EWHC 579 (admin)

General Medical Council v Meadow [2006] EWCA Civ 1390

Gillick v. West Norfolk and Wisbech Area Health Authority 1986 A. C. 112

Hussein v William Hill group [2004] EWHC 208

Toohey v Metropolitan Police Commissioner [1965] AC 595

R v Turner [1975] 1 All ER 70

America

Austin v American Association of Neurological Surgeons, 253 F 3d 967 (7th Circuit, 2001).

United states v. Jenkins, 420 u. s. 358 (1975)

Australia

Mustac v Medical Board [2004] WASCA 156

Re Noble [2002] PRDB (Vic) 6

Re Watson-Munro [2000] PRBD (Vic) 4

New Zealand

B v Attorney General [1997] NZFLR 550 (FC)

RIG v Chief Executive of the Ministry of Social Development HC
    Auckland CIV-2008-404-0033461
Lai v Chamberlains [2007] 2 NZLR 7

7

COMMUNICATION CONTEXTS
Social Presence in
Virtual Learning Environments

Online social presence (Krejins, Kirschner, Jochems & van Buuren, 2010) takes place with the recognition that much of social interaction, particularly in the contemporary e-enabled workplace, is not with others who are always immediately physically present, but with representations of others through the mediums of email, film, internet, blog, online forum, teleconferencing, and other technologies. Short, Williams and Christie (1976), the originators of social presence theory, defined social presence as "the degree of salience of the other person in a mediated interaction and the consequent salience of the interpersonal interaction" (p. 65). Entire relationships may be conducted through mediated technology and people increasingly rely on them as substitutes for F-t-F (face-to-face) interaction (Biocca & Harms, 2002, p. 7).

According to Kehrwald (2007, p. 504), there are a number of social-relational mechanisms which arise out of information provided by online social presence cues: commonality, trust, feelings of safety, respect, rapport, and interdependence. Commonality is the concept of mutuality or having something shared in common, and it involves the notion of "reciprocity" or exchange. It is also important to inspire feelings of safety and the creation of an environment which fosters feelings of trust and promotes interpersonal interaction. A safe environment creates a positive atmosphere where participants feel safe from negative behaviors. In the distance education context, it is important to distinguish between personal information and pedagogical information. The issue of trust and gaining the correspondants' confidence in the distance learning is essential for both pedagogical and interpersonal reasons. However, cognitive trust (based on the idea that others are reliable) may be more easily established than affective trust (involving reciprocation of emotion) in the online environment (Orviss & Lassiter, 2006, p. 168).

Introduction

With the proliferation of mobile phone and internet services and their adoption by large proportions of the global population, these virtual communication mediums have become "inseparable elements of emerging late-modern societal forms such as personal communities, network sociality and mobile sociality" (Petric, Petrovcic, & Vehovar, 2011, p. 117). Social presence in the online environment has been defined as "being with others" (Heeter, 1992), and the "level of awareness of the co-presence of another human, being or intelligence" (Biocca & Nowak, 2001). In virtual or Computer-Mediated-Communication (CMC), we experience others in terms of sensing and interacting with them, "not with their immediate embodiments of mind, i.e., physical bodies with their actual faces and voices, but with mediated embodiments of minds, representations made of pixels, ink, stone, paper, etc." (Biocca & Harms, 2002, pp. 4-5). Often we might simply imagine the other person with whom we are conversing, or even suspend our perception of them and respond simply to the textual communications that they make with us. However, smiling at emotions, laughing or crying in films, gaining an understanding of how another person thinks or feels, are all measures of online social presence. Depending on the nature of the interaction, these may be superficial or strong. Biocca and Harms (2002, p. 6) point out that people over millennia have used representations of others, be it from stone sculptures, wooden masks, or in virtual graphic characters for the purposes of transmission, storage, interface, mediated interaction, and decoration.

We may view such events as smiling at emoticons as relatively commonplace, but the question remains: how can these mediated, artificial, or cybernetic environments explain the production of social presence? Is it as simple as because there are human beings using (as in communicating through) these environments? How is it possible that you or I can be sitting in front of a computer screen made from thousands of tiny pixels of liquid crystals displayed in Lower Hutt,

New Zealand and gain a sense of shared embodiment and perceptual understanding with a colleague sitting in front of a computer screen of similar quality in, for example, Moscow, Russia?

The most common method of communication in the virtual world is the email. An email does not itself have a facial expression, interact with us (even if animated), or demand attention, but it nevertheless allows two or more people the ability to communicate across space and time. Technologies which enable social presence have increased in the contemporary world of communication, and so too have the number of individuals or interactions that a person has access to. Entire relationships may be conducted through mediated technology and people increasingly rely on them as substitutes for face-to-face (F-t-F) interaction (Biocca & Harms, 2002, p. 7). Consequently, the ability to distinguish differences in F-t-F and virtually mediated social presences requires a theoretical framework.

Clearly some mediums are better for some tasks, (email for short text-based conversations and digital video for contextualized descriptions), but these questions – the differences between responding to text or to a virtual representation of a physical body – remind us that online social presence may also be considered as a form of semiotics which has universal properties.

Definitions of Social Presence in e-Learning

The Networked Minds Theory of Social Presence attempts to provide this (Biocca & Harms, 2002, p. 7). At the basis of this theory is the relationship between two or more people, both with a sense of the "other" and of the other's sense of me, you, or us. This is tempered by the fact that in virtual communication, an individual might be aware of the other, but the other and the self may not be aware of an observer (Biocca & Harms, 2002, p. 18). This may be understood as the technology 'blind-spot' alluded to in the conceptualization of the Johari window (Luft, 1984, p. 60). This idea of a tactile 'distance' in communication (whether real or imagined) informs concepts of social presence. Indeed, the sense of 'being' at a 'distance' is a defining feature in terms of qualities of perception.

The primary definition of social presence is derived from Short, Williams and Christie (1976), who understand social presence as the extent to which a person is aware of another person in a technology-mediated setting. As Kehrwald (2007, p. 503) suggests, contemporary definitions of social presence include: the ability to perceive others in mediated interactions in virtual environments, the degree of proximity and tangibility of others in virtual communication, and the ability for self-projection in an online community.

But is social presence a latent characteristic of the communication medium, a property of perception in the participant, or a combination of the two? As Leong (2011, p. 8) points out, social presence is commonly measured through a semantic differential technique composed of a series of scales that measure perceived qualities of sociability: personal-impersonal, sensitive-insensitive, warm-cold, and sociable-unsociable. Steinmetz and Mussweiler (2011) suggest that ". . . the association of physical warmth and social similarity may well apply to non-social similarity perception, so that non-social objects seem more similar if physical temperature is higher" (p. 1026). The intriguing possibility is that this might apply also to perceived temperature, which corresponds to an emotional valency in the communicator, influencing a wide range of interpersonal experiences (Steinmetz and Mussweiler, 2011, p. 1028). However, these measurements are properties or *qualia* inherent to the perceiver. Tu and McIsaac (2002) conceive of three dimensions of social presence: (1) social context, (2) online communication, and (3) interactivity. The factors involved in these dimensions are given in Table 1 below (after Leong, 2011, p. 8):

Table 1. Showing factors in the three dimensions of social presence

(3) Interactivity	(1) Social Context	(2) Online Communication
CMC as pleasant	CMC as a social form	CMC conveys feelings and emotion
CMC as immediate	CMC as informal or casual way to communicate	Language used in CMC is stimulating
CMC as responsive	CMC as a form of personal communication	Language used in CMC is expressive
CMC as comfortable in use	CMC as a sensitive means of communicating with others	Language used in CMC is meaningful
CMC as familiar in use	CMC as easy and inclusive	Language used in CMC is easily understood

Biocca (1997) augmented the standard definition for social presence, stipulating that the minimum level of social presence occurs when CMC users perceive a form, behavior, or sensory experience that indicates the presence of another intelligence. Here he did not distinguish between biological and technological intelligence. However, Kehrwald (2007) does distinguish between biological and non-biological social presence. Indeed, he asks whether it makes sense to even conceive of a non-biological social presence. This definition is based on self/other relations and human or at least cybernetic identity capable of expressing personality (emotion, personal history and social awareness).

Kehrwald's (2007, p. 504) definition is thus: "Social presence is an individual's ability to demonstrate her [sic] state of being in a virtual environment and so signal her [sic] availability for interpersonal transactions." Kehrwald distinguishes between the production of social presence and the environment it is produced in (social presence is not in itself a quality of media), and also assigns social presence a performative quality. Thus to some extent, the degree of social presence is based on characteristics of the medium and the user's perception. However, Tu and McIsaac (2002) confirm Leong's view of the three dimensions of social presence – social context, online communication, and interactivity. They differentiate between interactivity and social presence, positing that awareness and influence must also be involved in social presence (Tu & McIsaac, 2002, p. 135).

Social presence may be diminished by low activity online. However, Gunawardena (1995) points out that communication failures often occur more at the social level than the technical level. Kehrwald (2007) emphasizes that interactions in virtually mediated technologies remain largely text-based, despite the increasing use of teleconferencing. Furthermore, Sharpe and Pawlyn (2009, p. 502) suggest that, "Moderated online asynchronous discussion board activity is still the commonest form of interaction online, whether in programmes [sic] which are fully online or blended." Sociability is limited by the relative perceptual 'leanness' of the textual medium. This leanness is defined by: a) lack of contextual information, b) psychological distance between actors and introduced media, c) imbalances in sender-receiver information due to lack of synchronous two-way interaction (2007, p. 502). Despite this, significant numbers of online participants report positive experiences in online learning, the transcending of time and space, and certain qualities of interactive exchanges surpassing experiences of other delivery modes, including F-t-F (Kehrwald, 2007, p. 502). Petric et al. (2011, p. 121) distinguish four elements in the use of interpersonal communication technologies capable of imparting social presence. These are: 1) Informational-cooperative use – giving and receiving information, transmitting knowledge, and learning, 2) Relational use – establishing and maintaining social relationships, giving and receiving social support, 3) Expressive use – communication acts that relate to the subjective world of personal experience, and 4) Strategic use – using interpersonal communication media to achieve personal goals and actions.

Based on telepresence research, Krejins et al. (2010) define social presence as a "degree of illusion that others appear to be 'real' physical persons in either an immediate (i.e. real-time/synchronous) or a delayed (i.e. time-deferred/asynchronous) communication episode" (p. 1). This tends to blur the distinction between biological presence and non-biological presence, emphasizing either the virtual or cybernetic qualities of social presence. However, as Bailenson et al. (2004) suggest, interpersonal distance in virtual mediums is similar to interpersonal distance behavior in physical environments; the personal space bubble is similar in size and shape.

Kehrwald (2007, p. 503) extends the idea of social network analysis (SNA) for understanding technologically supported social processes. SNA is characterized by relations (pathways for exchange between individual social actors) and ties (connections between social actors). According to SNA, relations and ties produce social systems. Relations pathways are characterized by content (e.g. information, electronic resources, various forms of support), direction (linear from me to you), or un-direction (through incidental interaction), and strength (e.g. frequency of interaction, emotional content, productivity, and ability to deal with uncertainty). Ties are also characterized by content, direction, and strength (Kehrwald, 2007, p. 503). Complex relations and ties can produce multiplexity (Kehrwald, 2007, p. 503).

Biocca and Harms (2007, p. 11) define mediated social presence as a property of people rather than of technologies, and as characterized by a phenomenal state produced by the virtual representation of another human being. The experience of social presence may be on a continuum from low-level awareness of co-presence to more intense perception of others' intentional states (Dennett, 1987). Perceptual awareness of the other is thus a minimum standard, but may also include awareness of the spatial co-presence of the other's body, and minimal attributes about the internal states of the other's identity and sentience. This awareness may increase towards more intense (and less synchronically comprehensible) recognition of the behavioral and psychological engagement of the other. It may be represented in the diagram below:

Figure 1. Levels and dimensions of social presence

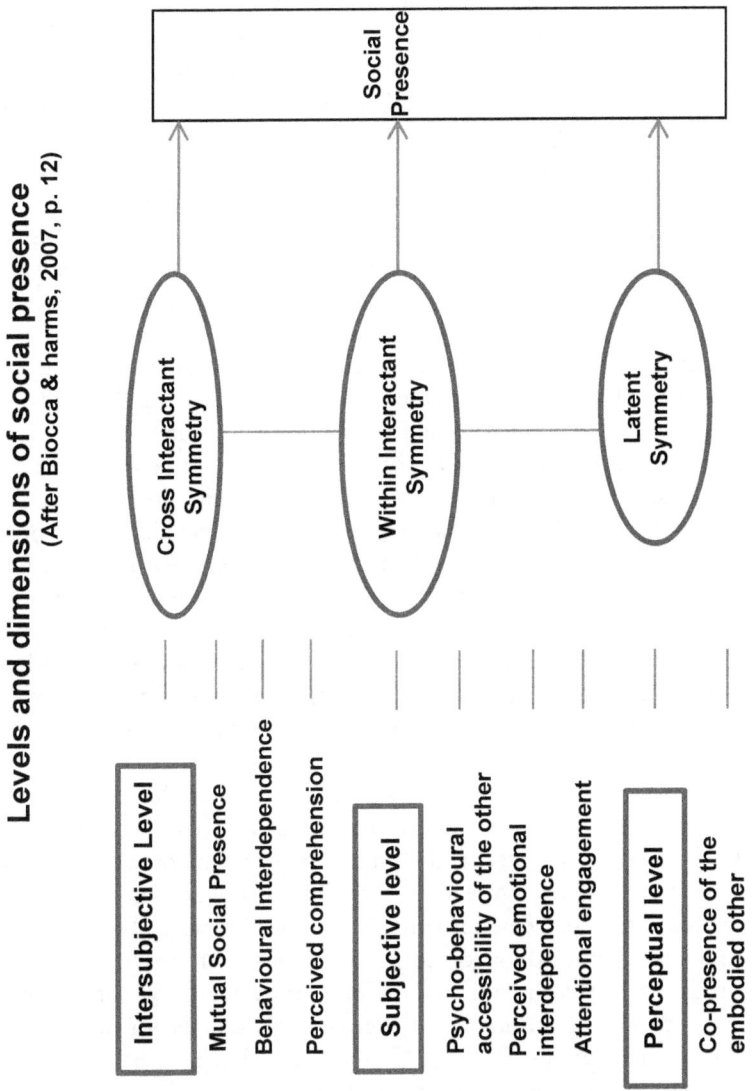

Figure 1 *Level and dimensions of social presence* above defines the causal factors of social presence, showing how an inter-subjective level involving mutual social presence, behavioral interdependence, and perceived comprehension results in cross-interactant symmetry in the creation of an online social presence. As Biocca and Harms state, "cross-interactant symmetry is the degree of symmetry or correlation between the user's (A) sense of social presence and their partner's (B) perception of the user's social presence" (2002, p. 29). A second subjective level of online communication involves perception of the psycho-behavioral accessibility of the other, attentional engagement and perceived emotional interdependence. This results in within-interactant symmetry in the creation of social presence.

As Biocca and Harms state, "within-interactant symmetry is the degree of symmetry or correlation between the user's (A) sense of social presence and their perception (A->B) of their partner's sense of social presence" (1997, pp. 28-30). These two levels are also paralleled by a perceptual level involving the recognition of the co-presence of the embodied other. This is understood as being 'latent symmetry' in so much as it is a basic level of comprehension of communicative understanding of the creation of social presence.

The types of mediums through which social presence may be articulated or experienced are manifold. They include: 1) collaborative work environments (mediated work interactions/growth in tele-communication infrastructure), 2) mobile and wireless telecommunication, 3) high band-width teleconferencing interfaces (tele-immersive F-t-F simulations), 4) agent-based e-commerce and help interfaces (intelligent agents that inhabit virtual environments, 5) speech interfaces, 6) 3D social virtual environments (Biocca & Harms, 2002, p. 3).

Biocca and Harms (2002) regard "co-presence of the embodied other" as an environmental affordance that is correlated with sentient beings. Mediated embodiments of sentient others almost always trigger social presence responses. These responses are even triggered by representations that the viewer knows to be false. The moment of perception of co-presence (whether derived from ink, pixels, or marble) is described as the "threshold moment" (Biocca, 1997, p. 13). Goffman (1963) derived two criteria for determining the moment of co-presence (whether biological or cybernetic): first, when individuals sense they are able to perceive others, and, secondly, the moment of perception when others perceive them (Biocca, 1997, p. 13).

In 3D environments vertical bilateral symmetry is a primary cue of co-presence of the other. According to Cui, Wang and Xu (2010), there is a division between "feature-based (actual, structural, and objective) recognitions of online social presence and perception-based recognitions (experiential and subjective)" (p. 37). However, bilateral symmetry can only be non-directly inferred from text-based communications. Attentional awareness is also a factor and the environmental subject must be perceived as biological and not an object. The Networked Minds Theory of Social Presence determined social presence as the extent to which the user feels whether he or she is in a "shared environment" of co-location (Biocca, 1997, p. 17). As Biocca (1997) suggests, ". . . it would appear that, social presence, defined as the sense of access to the other's minds, may involve some element of attentional allocation to the other as part of this shared attentional mechanism" (p. 18).

3D environmental measures of co-presence include awareness, eye fixation, proxemic behavior, and physiological responses. In the text-based environment, heightened social presence may be defined by description of self, self-other interaction, mimicry of facial expression, demonstration of intentional awareness, access to shared thinking, dispositional states of mind, access to shared affective states, decentring behaviors, sense of the intentional states of the other, referential engagement, reciprocity, motor mimicry, and recognition of phenomenal states (Biocca, 1997, pp. 21-25). There is also clearly an intentional reflexivity in the sense of the person being close enough to sense being perceived by the other (Goffman, 1963). Furthermore, the co-orientation model of interaction suggests that online social presence may be assessed by comparisons of subjective and inter-subjective responses to stimulus – a shared experience.

Social Presence and Collaborative Learning

As Joyner (2001) suggests, "lack of social presence and attrition of online learners may be connected" (p. 1). Furthermore, as Tu and McIsaac (2002, p. 30) observe, if social presence is low, social interaction and social learning cannot occur. Online social presence and social learning theory are thus related, and social presence enhances online social interaction. As Lui, Gomez, and Yen (2009, p. 165) observe social presence is also a significant predictor of both course retention and course grade. If social interaction is a prerequisite for collaborative learning and knowledge construction as Krejins et al.

(2010) suggest, then it also helps promote explanation and critical thinking. Thus social interaction as a component of social presence activates cognitive processes necessary for memory and understanding – it is consequently a vital component of the learning process (Krejins et al., 2010, p. 1). But social presence also has an affective component, the socio-emotional processes which affect group formation and dynamics and which may build cohesiveness and community (Krejins et al., 2010, p. 1). As Tu and McIsaac (2002, p. 132) point out, two central questions are: how does social presence theory apply to CMC systems, and how can social presence affect online learning?

However, Krejins et al. (2010) also argue that social presence theory is an example of technological determinism, the concept that technology also drives social structure and cultural values. Thus Krejin et al. are closer to Kerhrward in positing a cybernetic theory of social presence, or at least one which acknowledges social determinism. Tu and McIsaac (2002) add two other variables to this understanding of social presence theory: 1) system privacy – the operator's knowledge of the security of communication while using technology – known to self or others, and 2) privacy feelings – the perception of psychological security regardless of the actual security (p. 297).

However, for Garrison, Anderson and Archer (2000, 2001) social presence is one of three core elements of educational experience – the others being cognitive presence and teaching presence. As Cobb (2009) suggests, cognitive presence refers to the ability of participants in a community of learning to "[c]onstruct meaning through sustained communication" (p. 243). Teaching presence involves "designing and managing learning, providing subject matter expertise, and facilitation of active learning" (Cobb, 2009, p. 243). It is a means of enhancing social and cognitive presence to support teaching activities. Social presence is characterized by a further three factors: expression of emotion, open communication, and group cohesion (Cobb, 2009, p. 243). Social presence is viewed as critical to cognitive presence, and is characterized by a "qualitative difference between a collaborative community of inquiry and a simple process of downloading information" (Garrison et al., 2000, p. 96). Social presence is thus a characteristic of both learners and teachers.

Rogers and Lea (2005) also take a more technological view of social presence, distinguishing it from physical presence. The former implies the sender being physically located 'somewhere' in which the real medium is invisible to the receiver, whereas the latter prefers being in a form of communication in which the virtual medium itself appears to be a social entity (2005, p. 151). Although mediums low in social presence are seen as less social, nevertheless they may produce strong social influence under some circumstances (Rogers and Lea, 2005, p. 152). To some extent, in computer-mediated environments the social is equated with the interpersonal. The reduced social cues model assumes that "as (text-based) computer mediated environments do not allow for communication on non-verbal cues such as gestures or facial expressions, which impact on Face-to-Face interpersonal communication, these media are less social and therefore enable less social presence" (Rogers and Lea, 2005, p. 152). Consequently, the pedagogical implication is that for virtual environments to create social presence, teachers need to make frequent use of visual and audio social cues.

Characteristics of Social Presence in e-Learning

According to Leong (2011, p. 1), online social presence is related to student satisfaction but its impact is mediated by cognitive absorption. Furthermore Gunawardena and Zittle (1997) determined that social presence explained 60% of the variance in overall learner satisfaction. Pace (2004) suggests that there is a relationship between flow (which is derived from cognitive absorption) and telepresence. As Leong (2011) suggests, "when a person experiences flow, hours seem to transform into minutes while seconds may last for hours" (p. 9). Heeter (1992) describes telepresence as the sense of presence a person experiences when he or she is physically removed from the discussion (e.g. in an online forum). It has three basic components: personal presence, social presence, and environmental presence. Social presence is an antecedent to cognitive absorption and acts as a mediator variable between social presence and student satisfaction (Leong, 2011, p. 9). Unsurprisingly, interest in the subject matter may result in a higher quality online learning experience (Leong, 2011, p. 24). Directions for the instructors of online courses include facilitating interest in the courses by emphasizing relevance in students' daily lives and pursuing interactive instructional activities (Leong, 2011, p. 24).

The relationship between online communication and student satisfaction is indicated in Figure 2 *Contributing factors to student online satisfaction* below.

Figure 2. Contributing factors to online student satisfaction (After Leong, 2011, p. 11)

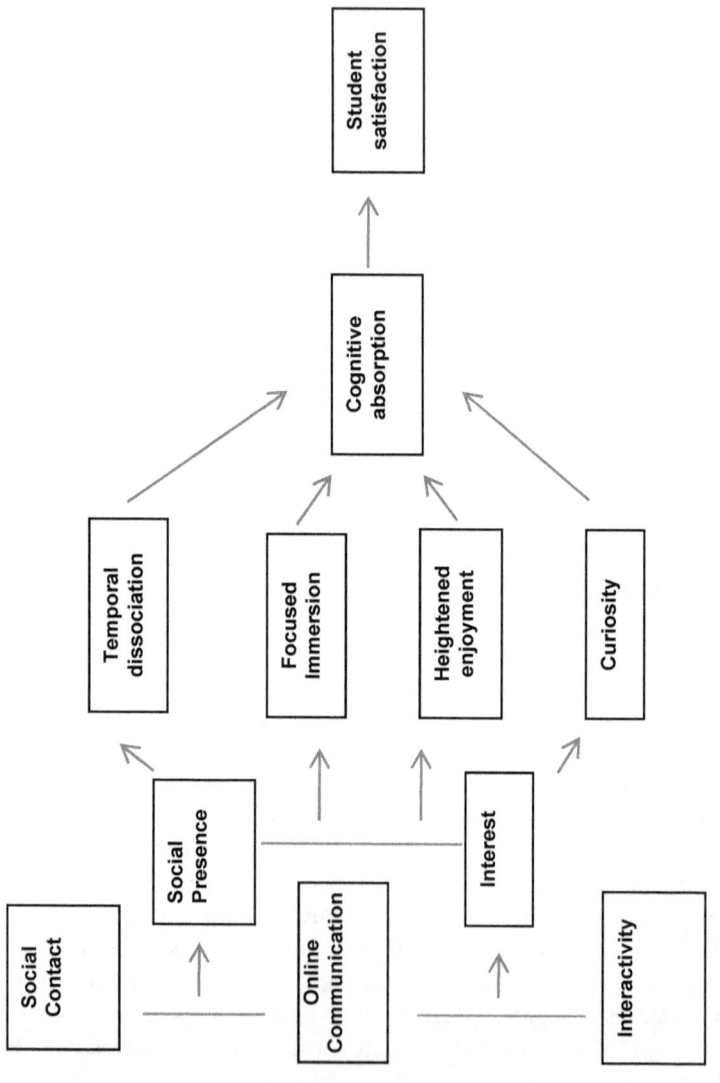

As Caspi and Blau (2008, p. 323) suggest there are correlations in online group communication between self-projection, perception of others, group identification, and perceived learning. Absent from these correlations is the users rating of the subjective quality of the virtual medium. Caspi and Blau (2008) suggest that social presence may stimulate learning by setting a convenient climate for it. Furthermore, it may stimulate the socio-emotional source of learning and potentially also the cognitive source of learning. Establishing social presence is of course challenging in a text-based environment.

Caspi and Blau (2008, p. 324) argue that social presence has three main characteristics. These are: 1) it is a medium that allows transmissions of social cues essential to perceive another learner as real; 2) It has the ability for the learner to project him or herself emotionally and cognitively as a real person in an online community; 3) it contains the reflection of a level of identification with an online learning group. The perception of immediacy is also a factor of social presence, described by Mehrabian (1969) as "those communication behaviors that enhance closeness to and nonverbal interaction with another" (p. 203). More distant interactions are the result of conversations with less social clues.

Caspi and Blau (2008) also argue that while communication via computer is seen as less phenomenologically interactive than F-t-F communication, it is nevertheless more personal, focused and immediate. They argue that text-based interactions allow the learners to build composite images of the person with whom they are interacting, in effect a live image "*despite* and perhaps *because* of the limitation imposed by the medium" (Caspi & Blau, 2008, p. 339). It is certain that different perceptions of online social presence may alter the ways students present themselves and perceive others. It is possible also that use of CMC may heighten users' sensitivities to social communication clues, and this may also enhance self-projection in an online group (Caspi & Blau, 2008, p. 339). In an online environment which presents little social clues, what remains are social category clues. Correspondingly, in this situation the social identity of the group will determine the behavioral norms of communicators (Caspi & Blau, 2008, p. 339). Surprisingly, the "impersonality" of the medium has a weak correlation to learning variables.

Social presence thus spans different classes of technology. It can be found in a variety of mediated interactions. It applies to human and cybernetic interactions. It concerns 'real' or 'illusory' social interactions. However, the SIDE (Social Identity Model of Deindividuation Effects) obviates the necessity for interpersonal interaction in social presence since the lack of non-verbal or bodily cues in CMC may increase rather than decrease social presence in group contexts (Rogers and Lea, 2005, p. 152). Underlying the SIDE model is the belief that individuals have multiple layers of self, including personal identity and a range of possible social identities. The salience of a relevant social identity can thus be increased by contexts in which transfer of personal information is limited. Instead of individuating or conceiving in interpersonal terms of self versus other, relations communication by people in groups is more likely to be in terms of shared similarity than differences – a social rather than personal identity. This depersonalized shared social identity has an increased influence on behaviors (Rogers and Lea, 2005, p. 153).

As Mykota and Duncan (2007) state, "he genesis of social presence lies in the conceptualization from social psychology of immediacy and intimacy surrounding F-t-F communication" (p. 159). As Kehrwald (2007, p. 509) suggests, social presence is important because it is not possible to build a community of learners without it. As well as being socially inclined, it is also impossible to engage with pedagogy in an agentively (and thus politically neutral) way; values, beliefs, and conceptualizations of self are all implicated in learning (Kehrwald, 2007, p. 509). Perhaps because of this it is recommended that online teaching and learning practice include the development of strong social presence by the facilitator and in online instructional designs which require interpersonal and group interaction (Kehrwald, 2007, p. 510). Thus social presence plays a prominent role in text-based online learning environments as it gives individual information about social actors and indicates the relative strength of relations between online agents. Relational and content information is critical for building ties between actors for group activities, thus enhancing the quality of learning collaborations (Kehrwald, 2007, p. 510).

Space perception and transactional distance are also factors in the creation of social presence. Perceptions of space are somewhat subjective and may influence online participation. Transactional distance describes the sense of psychological and communications distance to be surmounted online. Although a source of exchange and creativity, it is also a space of potential misunderstandings between instructors and learners who may be separated by both space and time (Moore, 1993, p. 22). Stein and Wanstreet (2003, p. 194) describe transactional distance as a feature of course design. High structure and low dialogue result in larger transactional distance and more emphasis on the learner to be autonomous. Transactional distance reduces in courses with frequent dialogue and low structure because learners receive ongoing communication and instruction, which may be modified.

Hypotheses and Correlations about Social Presence in e-Learning

One way of conceptualizing social presence is by the ability to transcend the potential of communication via the virtual medium, in order to practice or actually perceive the other person making it. Setting aside notions of the limitations of the virtual medium, to overcome constraint is part of the agentive activity of online social presence (Caspi & Blau, 2008, p. 325). Perception of online social presence is also related to what has been happening immediately prior, because if students perceive an online environment replete with social presence, they will also project more of themselves into the fora (Caspi & Blau, 2008, p. 325).

As Richardson and Swan (2003, p. 77) argue, students' perceptions of social presence may be related to their perceived learning and instructor. According to Caspi and Blau (2008, p. 328), there are four main inferences which may be drawn about the perceived learning through social presence in online mediums. First, social presence is a necessary precursor to learning and may inspire a high level of perceived learning if an inference is made on the socio-emotive aspects of learning. Secondly, while social presence and perceived learning positively correlate, there is no direct causal relation as social presence is seen as a by-product of online learning and social interaction; a distinction needs to be made between the actual learning experience and the accompanying experience. Third, it is possible that social presence may detract from perceived learning. Fourth, it is possible that social presence and perceived learning are not related since perceived learning is the result of cognitive rather than emotive processes. Furthermore Caspi and Blau (2008, p. 338) found that the impersonality of the virtual medium is not related to concepts of group identification, nor does it correlate with the perception of the other. As Rettie (2003, p. 3) suggests, social presence and connectedness are related but not synonymous. Social presence involves a perception of the medium whereas connectedness is an affective experience inspired by the medium.

The Creation of Social Presence

Sallnäs (2005, p. 435) identifies two main factors in the establishment of presence in virtual environments. The first is involvement, ". . .

the degree of significance or meaning that individuals attach to stimuli, activities or events" (p. 435). The second is immersion, ". . . a psychological state characterized by perceiving oneself to be enveloped by, included in, and interacting with an environment that provides a continuous stream of stimuli and experiences" (Sallinas, 2005, p. 435).

Suler (2003) defines four factors in perceiving online presence: sensory stimulation from the environment, change in the environment, interactivity with the environment, and the degree of familiarity. Trust includes being able to be trusted and behaving in a trustworthy way. There are six factors involved in the creation of online trust: 1) confidence in the other party, 2) comfort in interacting with others, 3) courage with 'having a go' and participating, 4) respect is a positive relational condition between people, 5) rapport requires mutuality, trust and respect, 6) interdependence refers to the concept that learners are mutually supportive (Kehrwald, 2007, p. 506). Relations and ties form between individuals online in relation to interactions and transactions between individuals (Kehrwald, 2007, p. 507).

As Kehrwald (2007, p. 507) suggests there are five main types of the development of relations between online parties. These are summarized in the table below:

Table 2. The progressive development of relations in online group work

Progression of relations	Relational mechanisms at work
Me-other relations	Empathy Respect Admiration
Mutuality	Commonality Connection Like-mindedness
Feelings of safety	Freedom from risk Comfort with others Confidence in others
Trust	Trustworthiness Trusting behaviors Willingness to put oneself at risk
Production	Group cohesion Rapport Interdependence

Mykota and Duncan (2007, p. 160) have suggested that if social presence is extremely low it may result in learner frustration because of diminished quality of interpersonal interactions. Social presence thus inspires cognitive presence. However, if there are affective as well as cognitive goals for the education process, participants will find group interaction enjoyable and fulfilling; thus social presence is a contributor to the learning process (Mykota & Duncan, 2007, p. 160). Rourke, Anderson, Garrison, and Archer (1999) define three response categories in online communication: affective indicators (values, beliefs, feelings, emotions), cohesive indicators (group commitment, mutuality), and interactive indicators (the creation of social meaning). Mykota and Duncan (2007, p. 166) assert that social presence is formed by sharing experiences online by participating in chats, posting discussion messages, providing a personal biography, through instructional design strategies, and development of collaborative assignments.

According to Leong (2011, p. 5), instructor variables such as communication, feedback, preparation, content knowledge, teaching methods, accessibility and interactivity are all important in the creation of social presence in the learning experience. Tu and McIsaac (2002) and Aragon (2003) assert that the following are key factors in the preparation of online classes:

- Timely response to messages

- Use of stylistic communication types/humor, emotions

- Casual conversations/personal stories

- Appropriate message register and length

- Planning, creativity, decision-making, social tasks

- Appropriate communication group size

- Inclusion of student profiles.

There is also evidence that the appeal of social presence in enhancing online engagement also works in the commercial setting. As Hassanein and Head (2005/6) suggest, it is about "increased levels of social presence through socially-rich descriptions and pictures positively impacts attitudinal antecedents" (p. 1). Clearly social presence is a facilitator of engagement in all online environments.

However, social presence, teaching presence and cognitive presence are three main factors in teaching and learning in online environments. According to Vaughan and Garrison (2006), an online instructor can establish social presence by using emoticons, openly expressing emotions, trust, judgment, and by encouraging collaboration between learners. They are summarized by Vaughan and Garrison (2006) in the table below:

Factors	Categories	Examples
Social presence	Affective expression	Emoticons
	Open communication	Express trust and judgment
	Group cohesion	Encourage collaboration
Teaching presence	Design and organization	Defining content and activities
	Facilitating discourse	Sharing meaning
	Direct instruction	Focusing discussion
Cognitive presence	Triggering event	Sense of puzzlement
	Exploration	Information exchange
	Integration	Connecting ideas
	Resolution	Applying new ideas

Social presence is described as the ability of learners and instructors to project themselves in cyberspace as real people. Teaching presence involves the facilitation and stimulation of cognitive and social modes of enquiry for the purposes of learning. Cognitive presence involves the ability of learners to construct and accurately invent meaning through discursive, creative or critical enquiry within a community of learners.

Psychological Factors Influencing Inhibition/Disinhibition in Online Learning

The quality of human communication is defined as much by the subtle exchange of emotional inflections to tinge words with meaning as it is by the symbolic content of communication, a process called prosody. The inability to distinguish emotion in speech is a condition called aphrosia. Lewis, Amini and Lannon (2001) have gone so far as to suggest that "millions of people experience daily aphrosia in their email" (p. 59). Suler (2003) identifies nine psychological factors influencing inhibition/disinhibition in e-learning. These are summarized in the table below.

Table 3. Psychological experiences influencing inhibition/disinhibition in online learning

Online disinhibition effect	People say and do things in virtual environments that they wouldn't do in f-t-f world. This leads to: Benign disinhibition Toxic disinhibition
Dissociative anonymity (you don't know me)	People can't easily tell who you are in virtual environments. People can separate actions from real world identity.
Invisibility (you can't see me)	Power to be concealed overlaps with anonymity (concealment of identity). Physical invisibility amplifies the disinhibition effect.
Asynchronicity (see you later)	People don't have to interact with one-another in real time – suspending time before reply produces either a disinhibiting effect or in e-learning, an accuracy correlation.
Solipsistic introjection	Absence of f-t-f cues can inspire feeling of merging minds
Dissociative imagination	Creation of a 'dream-world' persona – separate from demands and responsibilities of real life

Minimizing authority (we're equals)	Although impact may be lessened – everybody on internet has opportunity to voice him or herself.
Personality variables	Strength of underlying feelings and personalities influences tendencies toward inhibition or expression online
True Self	Exploration of concepts of self online: Personal and cultural values Inhibiting/disinhibiting self Compromise formations Self constellations across media Altering self-boundary

The psychological exploration of self online is complex. Within the concept of the online self, the notion of 'self-boundary' is a strong factor in online communication and e-learning. As Suler (2003) suggests, a self-boundary is a sense of 'what is me and what is not me.' It may be a continuum between inhibition and disinhibition. The self-boundary describes a flexible perimeter which makes distinctions between thoughts, feelings, and memories apparent not only in terms of what exists outside that field but also within other people. Suler (2003) identifies a plethora of factors that contribute to a self-boundary: a) awareness of having a physical body, b) perception via senses of the outside world, c) being able to make a psychological distinction between what I know and what others know about me, d) sensation of the physical/psychological self moving in a space/time continuum of past present future. These can become disrupted by cyber-space – not only in terms of disembodiment (the physical body no-longer plays a crucial role), but what is known about self and other becomes unclear Space and time can become distorted with synchronous and asynchronous communication. This may destabilize boundaries whereby the distinction between the inner and outer me is not clear. In these situations of diffused 'self and other' boundaries, 'primary process thinking' may take over in the interpretation of online social presence.

Conclusion

As Biocca (1997) suggest, "humans are the only species that engage in sustained and prolonged interaction with representations of others" (p. 14). Social presence is concerned with defining the digital 'self,' 'group' or 'other.' As Charnock (2012, p. 12) says, this is a complex mosaic of habit, subconscious acts of omission and commission, and premeditated presentations. In this sense the 'digital you' of social presence takes on both mythical and cybernetic characteristics being "more than the sum of the breadcrumbs" left behind in travels through cyberspace (Charnock, 2010, p. 12). Rather, it entails a notion of mind and person which picks out "open-ended systems" capable of including "non-biological props and aids" as parts of themselves (Clark, 2003, pp. 9-10). Furthermore, social presence is an extension of the mind-body-scaffolding problem in biology which combines with technology to produce perceptions, emotions, and cognitions in a supportive environment (Clark, 2003, p. 11). Social presence also takes place in a complex psychological context, which may inhibit, amplify, or extend personality and behavioral expressions beyond those associated with the F-t-F environment.

Acknowledgments: My thanks are due to Dr Polly Kobeleva for comments and suggestions.

References

Aragon, S. R. (2003). 'Creating social presence in online environments'. *New Directions for Adult and Continuing Education, 100*, 57-68.

Bailenson, J. N., Aharoni, E., Beall, A. C., Guadagno, R., Dimov, A., & Blascovich, J. (2004). 'Comparing behavioral and self-report measures of embodied agents'. *Social Presence in Immersive Virtual Environments*. Retrieved from: http://www.stanford.edu/~bailenso/papers/presconf.pdf

Biocca, F. (1997). 'The cyborg's dilemma: Embodiment in virtual environments'. *Journal of Computer-Mediated Communication 3*(2). Retrieved from: http://www.ascusc.org/jcmc/vol3/issue2/biocca2.html

Biocca, F., & Harms, C. (2002). 'Defining and measuring social presence: Contribution to the networked minds and theory and measure'. *Media Interface and Networked Design (M.I.N.D.) Labs.*

Retrieved from: http://www.temple.edu/ispr/prev_conferen ces/proceedings/2002/Final%20papers/Biocca%20and%20Har ms.pdf

Biocca, F., & Nowak, K. (2001). Plugging your body into the tele-communication system: Mediated embodiment, media interfaces, and social virtual environments. In C. Lin & D. Atkin (Eds.). *Communication technology and society* (pp. 407-447). Waverly Hill, VI: Hampton Press.

Caspi, A., & Blau, I. (2008). 'Social presence in online discussion groups: Testing three conceptions and their relations to per-ceived learning'. *Sociology, Psychology, Education, 11*, 323-346.

Charnock, E. (2010). *E-habits: What you must do to optimize your profes-sional digital presence?* New York: McGraw Hill.

Clark, A. (2003). *Natural-born cyborgs. Minds, technologies, and the future of human intelligence.* New York: Oxford University Press.

Cobb, S. C. (2009). 'Social presence and online learning: A current view from a research perspective'. *Journal of Interactive Online Learning, 8*(3), 241-254.

Cui, N., Wang, T., & Xu, S. (2010). 'The influence of social presence on consumers' perceptions of the interactivity of web sites'. *Jour-nal of Interactive Advertising, 11* (1), 36-49.

Dennett, D. (1987). *The intentional stance.* Cambridge, MA: MIT Press/Bradford Books.

Garrison, D. R., Anderson, T., & Archer, W. (2000). 'Critical inquiry in a text-based environment: Computer conferencing in higher education'. *The Internet and Higher Education, 2*, 87-105.

Garrison, D. R., Anderson, T., & Archer, W. (2001). 'Critical think-ing, cognitive presence, and computer conferencing in distance education'. *American Journal of Distance Education, 15* (1), 17-23.

Goffman, E. (1963). *Behavior in public places: Notes on the social organiza-tion of gatherings.* New York: The Free Press.

Gunawardena, C. N. (1995). 'Social presence theory and implications for interactive and collaborative learning in computer confer-ences'. *International Journal of Educational Telecommunications, 1*(2/3), 147-166.

Gunawardena, C. N., & Zittle, F. J. (1997). 'Social presence as a pre-dictor of satisfaction with a computer-mediated conferencing environment'. *American Journal of Distance Education, 11*, 8-26.

Hassanein, K., & Head, M. (2005/6). 'The impact of infusing social presence in the web interface: An investigation across different products'. *International Journal of Electronic Commerce* (IJEC), *10*(2), 31-55.

Heeter, C. (1992). 'Being there: The subjective experience of presence'. *Presence: Teleoperators & Virtual Environments, 1*(2), 262-271. Retrieved from: http://mitpress.mit.edu/journal-home.tcl?issn =10547460

Joyner, C. C. (2009). Social presence in the asynchronous online learning environment: Doctoral learners' lived experiences. (Doctoral Dissertation). Retrieved from: http://www.grin.com/ en/doc/230296/social-presence-in-the-asynchronous-online-learning-environment-doctoral

Liu, S. Y., Gomez, J., & Yen, C-J., (2009). 'Community college online course retention and final grade: Predictability of social presence'. *Journal of Interactive Online Learning, 8*(2), 165-182.

Kehrwald, B. (2007). 'The ties that bind: Social presence, relations and productive collaboration in online learning environments'. *Proceedings Ascilite Singapore 2007.*

Krejins, K., Kirschner, P. A., Jochems, W., & van Buuren, H. (2010). 'Measuring perceived social presence in distributed learning groups'. Retrieved from: http://celstec.org/content/measuring-perceived-social-presence-distributed-learning-groups

Leong, P. (2011). 'Role of social presence and cognitive absorption in online learning environments'. *Distance Education, 32*(1), 5-28.

Lewis, T., Amini, F., Lannon, R. (2001). *A General Theory of Love*. New York: Vintage Books.

Luft, J. (1984). *An introduction to group dynamics*. Mountain View, CA: Mayfield Publishing Company.

Mehrabian, A. (1969). 'Some referents and measures of nonverbal behavior'. *Behavior Research Methods and Instrumentation*, 54(2), 205-207.

Moore, M. G. (1993). Theory of transactional distance. In D. Keegan (Ed.), *Theoretical principles of distance education* (pp. 22-38). New York: Routledge.

Mykota, D., & Duncan, R. (2007). 'Learner characteristics as predictors of online social presence. *Canadian Journal of Education', 30*(1), 157-170.

Orvis, K. L., & Lassiter, A. R. L. (2006). 'Computer-supported collaborative learning: The role of the instructor. In S. P. Ferris & S. H. Godar (Eds.), *Teaching and learning with virtual teams* (pp. 158-179). Hershey, PA: Idea Group.

Pace, S. (2004). 'A grounded theory of the flow experiences of web users'. *International Journal of Human-Computer Studies, 60*(3), 327-363.

Petric, G., Petrovcic, A., Vehovar, V. (2011). 'Social uses of interpersonal communication technologies in a complex media environment', *European Journal of Communication, 26*, 116-132.

Rettie, R. (2003). 'Connectedness, awareness and social presence'. *6th Annual International Workshop on Presence;* 6-8 October 2003, Aalborg, Denmark. Retrieved from: http://eprints.kingston.ac.uk/2106/1/Rettie.pdf

Richardson, J. C. & Swan, K. (2003). 'Examining social presence in online courses in relation to student's perceived learning and satisfaction'. *Journal of Asynchronous Learning Networks, 1* (7), 68-88.

Rogers, P., & Lea, M. (2005). 'Social presence in distributed group environments: The role of social identity'. *Behavior & Information Technology, 24*(2), 151-158.

Rourke, L., Anderson, T., Garrison, D. R., & Archer, W. (1999). 'Assessing social presence in asynchronous text-based computer conferencing'. *Journal of Distance Education, 14*(2), 50-71.

Sallnäs, E-L. (2005). 'Effects of communication mode on social presence, virtual presence, and performance in collaborative virtual Environments'. *Presence, 14* (4), 434- 449.

Sharpe, R., & Pawlyn, J. (2009). The role of the tutor in blended E-Learning: Experiences from interprofessional education. In Donnelly, R., & McSweeney, F. (Eds.), *Applied E-Learning and E-Teaching in Higher Education.* (pp. 18-34). doi:10.4018/978-1-59904-814-7.choo2

Short, J.A., Williams, E., & Christie, B. (1976). *The social psychology of telecommunications.* New York: John Wiley & Sons.

Stein, D. S., & Wanstreet, C. E. (2003). 'Role of social presence, choice of online or F-t-F group format, and satisfaction with perceived knowledge gained in a distance learning environment'. *2003 Midwest Research to Practice Conference in Adult, Continuing, and Community Education.* Retrieved from: http://www.alumni-osu.org/midwest/midwest%20papers/Stein%20&%20Wanstreet--Done.pdf

Steinmetz, J., & Mussweiler, T. (2011). 'Breaking the ice: How physical warmth shapes social comparison consequences'. *Journal of Experimental Social Psychology, 47*, 1025-1028.

Suler, J. (2003). 'Presence in Cyberspace'. Retrieved from: http://users.rider.edu/~suler/psycyber/presence.html

Suler, J. (2004). 'The Psychology of Cyberspace'. Retrieved from: http://users.rider.edu/~suler/psycyber/disinhibit.html

Suler, J. (2005). 'The Basic Psychological Features of Cyberspace'. Retrieved from: http://users.rider.edu/~suler/psycyber/basic feat.html

Tu, C-H., & McIsaac, M. (2002). 'The relationship of social presence and interaction in online classes'. *The American Journal of Distance education*, *16*(3), 131-150.

Vaughan, N., & Garrison, D. R. (2006). 'How blended learning can support a faculty development community of inquiry'. *Journal of Asynchronous Learning Networks*, *10* (4), 139-152.

8

UNDERSTANDING AMBIGUITY TOLERATION IN BUSINESS COMMUNICATION

This chapter explores the types of ambiguity that may be encountered in business communication, and their philosophical and language-based origins. It discusses the dependence of the business environment on language context, ambiguity in organizational climate, and the ethical uses of ambiguity. This chapter does the following: outlines the usefulness of ambiguity in strategic organizational communication, explores strategies of ambiguity in crisis and risk management, comments on ambiguity in organizational mergers and acquisitions, and explains ways in which ambiguity can both help and hinder business communication. Finally, the chapter explores leadership qualities for ambiguous situations.

Introduction

Communication theorists regard the reduction of uncertainty as a key factor in business communication (Salazaar, 1996, Dwyer 2009). Certainly ambiguity in the business environment has positive and negative qualities. Everyday positive features of ambiguity in business language include the use of metaphor to explain, promote cohesion, and provide inspiration. However, this may be contrasted with negotiating and contracting, in which the elimination of uncertainty is a necessary condition before forming any business agreement. Ambiguity in the workplace will always exist when there are two or more people engaging in symbolic interaction to exchange information (Mead, 2003). The sources of ambiguity are multiple. Ambiguity can result from indecision, an intention to mean several things at once, or the fact that a statement may have different meanings. For Eisenberg (1984), ambiguity is engendered through "detailed literal language as well as through imprecise, figurative language" (p. 230). However, ambiguity itself is ambiguous. Ambiguity of meaning may be totally independent of perceived ambiguity, which is a psychological variable, and may be different again from the ethical use of "strategic ambiguity" within an organization (1984, p. 230). As Eisenberg (1984) suggests, "ambiguity occurs in an organization when there is no clear interpretation of a phenomenon or set of events. . . ambiguity can exist within the organization as a whole as well as within individuals own cultural experience" (p. 259).

Participant Ambiguity in Language

Kahn, Wolfe, Quinn and Snoek (1964) define ambiguity as "the lack of clear, consistent information" (p. 23). Ambiguity may occur in the location and mobilization of symbolic communication, or simply, in the expression of language. Ambiguity occurs in the spaces of meaning which are not fixed or in which words may "gesture to things which are not fully understandable" (Komesaroff, 2005, p. 632), for example, ideas, emotions and experiences. Ambiguity occurs where language attempts to define meanings that are indirectly approached or may be multivalent (meaning several things at once). Ambiguity may be generated from inside language, not as "part of a formal, logical deduction involving an interlocutor but with a singularity located outside the subject of exchange" (Komesaroff, 2005, p. 633).

Ambiguity differs from deception in so much that ambiguous communication has the potential, a) to be either right or wrong, b) to be neither right or wrong, or, c) be simply metaphorical; in contrast, "deceptive language" is accomplished by falsification, concealment, or equivocation (Buller & Burgoon, 2003, p. 99). Ambiguous communicators do not necessarily intend to deceive, rather to imagine how things might be. However, deceptive communicators can usually be characterized by uncertainty and vagueness, non-immediacy, reticence, withdrawal, disassociation, or "image and relationship-protecting behavior" (Buller & Burgoon, 2003, p. 100). Nevertheless, people also may or may not exhibit these qualities in the normal course of communication.

Philosophically, constructionists argue that there is no purely objective reality to describe (that there are no objective world traits universally apparent) that are not mediated by language. Accordingly, knowledge is relational and interdependent, and new knowledge is discovered by extending existing knowledge. The use of ambiguity is a language device that may unlock new understandings (Eisenberg, 1984, p. 229). However, a linguistic relativist position does not consider ambiguity to be a special problem of language, suggesting instead that meanings are constituted by individuals, not inherent in discourse itself. The interactional view of communication proposed by Watzlawick and Weakland (1977) argues that all action is potentially communicative and the context is a key factor in determining meaning.

Empson (1996, p. 48) suggests there are three possible dimensions in language use along which ambiguities may occur. These are the degree of logical grammatical disorder, the degree to which the apprehension of the ambiguity must be conscious, and the degree of psychological complexity concerned. In his famous book of literary criticism, *7 Types of Ambiguity*, Empson defines the kinds of ambiguity in language use (summarized here):

1. Ambiguity by difference – a word may have several distinct meanings or several meanings may be interconnected, e.g. 'a sitting duck.'

2. Ambiguity by reduction – two or more word meanings may be resolved into one, e.g. 'point-blank.'

3. Ambiguity by definition – two ideas which are connected only by being relevant in context, can be given simultaneously in one word, e.g. 'a paradox.'

4. Ambiguity by distinct creation — two or more meanings of a statement do not agree amongst themselves but combine to produce a more complicated statement, e.g. 'your check has bounced.'

5. Ambiguities of transition – an idea occurs which applies to nothing exactly but lies halfway between two meanings, e.g. 'meet me half-way.'

6. Ambiguity of tautology – the reader is forced to invent intended understandings, e.g. 'leave before you go.'

7. Ambiguities by division – two meanings of the word, are the two opposite meanings defined by the context, e.g. 'she had a break.'

The implication is that language use is inherently complex in terms of the meaning it creates. As post-structuralists recognize,

Derrida's theory of linguistic trace adds further complexity to Empson's understanding of ambiguity in language use by showing that language meaning may be inherently shifting and ultimately indeterminate (Bradley, 2008, p. 7). However, the use of ambiguity in communication may be both inadvertent and advertent. Therefore, an objectivity test may apply in which language use considered subjectively ambiguous can be assessed in terms of the meaning that a 'reasonable person' would ascribe it. This may work well in print, however, language ambiguity in organizations that have both high and low context protocols is that it is frequently not enough to dismiss the ambiguous language as 'something' that could 'mean anything,' but rather better to think of it as a conceptual 'seed' of meaning that could grow in many directions. This raises the question of what makes selective attention worthwhile. Mostly this is a factor of varying attention span and shifting focus as much as it may be of rigorous analysis.

The Dependence of Business Environment on Language Context

Businesses depend on effective written and spoken communication for operational efficiency. If communication is poor, productivity decreases and labor and supervisory costs increase. If communication is clear and communication channels are designed efficiently, productivity increases (Cheney, Chistensen, Thoger, Ganesh, & Shiv, 2004). Similarly, the desire to reduce ambiguity in interpersonal communicative interactions is usually motivated by three elements: a) anticipation of future interaction, b) incentive value, or c) deviance (Berger, as cited in Griffin, 2003, p. 143). Even company morale may be linked to poor or effective communication and consequently company image (Piotrowski, 2005). The quality and uses of ambiguity in an organization may be complex and to some extent counterintuitive. For example, in order to achieve change an organization may need a comparatively high degree of complexity and a correspondingly high degree of precision, but in some situations, it is more efficient to be "ambiguous and more lowly-structured" in order to keep options open to adapt to change (Menz, 1999, p. 101).

Ambiguity in business organizations is of course not only limited to communication. The relative clarity or ambiguity of communication within a business organization can also be affected by the very nature of business itself. Kahneman and Tversky's (1979) prospect theory aims to describe how decision-making is informed by choices between risk-laden alternatives and known probabilities. These are implicitly or explicitly the decisions at the basis of much market-based behavior. Relatedly, expected utility theory explains how the price of a good or service will be related to marginal utility (its perceived value relative to other goods or services) (Bernoulli, 1738/1954). Put simply, efficient communication in business is a necessary but not sufficient quality.

At the basis of all communication practices are technology, relationships, and the use of language. Language is essential to human communication, which includes communication in the business context. As Piotrowski (2005) relates, communication issues can become compounded exponentially in the expedition of work. For example, "30% of letters and memos in industry and government do nothing more than seek clarification of earlier correspondence or respond to the request for clarification" (Piotrowski, 2005, p. 1). At the root of poor communication is ambiguity, which is the discursive understanding of 'indeterminacy' or being in a state of not knowing or 'misunderstanding'. However, written and spoken ambiguity may also bring positive benefits to communication, depending on what context they are used in.

Communication is ambiguous because people are complex, will seldom be thinking the same thoughts, may have divergent views, and because of cultural differences, no two perspectives are exactly alike. Furthermore, human interactions may be infinitely complex or very simple. Wallerstein (2004) questions whether we can we measure the impact of the perceiver on the perception, the measurer on the measurement? Something taken at face value by one person may be deeply significant to another (Wallerstein, 2004). This is the Heisenberg uncertainty principle writ large. Furthermore, he asks, "how can we get beyond both the false view that an observer can be neutral and the not very helpful suggestion that observers bring their biases to their perceptions?" (Wallerstein, 2004, p. 26). Clearly Wallerstein is skeptical of the ability for people to step outside their subjective views. However, in the everyday working context, this can usually be managed by the mediation of the work-task itself following from communication.

DeVito, O'Rouke and O'Neill (2000) suggest that uncertainty and ambiguity "increase with larger cultural differences" (p. 107). Contexts of ambiguity in communication may also exhibit intercultural differences. The basic distinction between 'high-context' cultures that place emphasis on personal relationships, local knowledge, and oral agreements and in which information may be known by all participants but not explicitly stated, and 'low context' cultures in which information is explicitly stated and often written down, also reflects differences in ambiguity toleration. As DeVito et al. (2000) state:

> To a member of a low context culture, what is omitted creates ambiguity. To this person ambiguity is simply something that will be eliminated by explicit and direct communication. For a member of a high-context culture, ambiguity is something to be avoided – it is a sign that the interpersonal and social interactions have not proved sufficient to establish a shared base of information. (p. 104).

Some workplaces also have a combination of high and low context protocols which can lead to conflict and interference in upward, downward, and horizontal communication. Communication is also a factor of people's motivations, generally understood to be money, sex, success, social status, power, emotional security, self-image, family welfare, national welfare, or getting through the day. Motivations are inherently complex and may be ambiguous or under-realized influences on many people's behaviors, self-fashioning, and perceptions of other's. No two people's motivations and thus communication responses to a given situation will be exactly alike. Unless communication is fundamental and basic, no two people will have exactly the same motivations for communicating about any one issue. Sunnafrank (1986) suggests that people are inclined to seek information not necessarily to reduce uncertainty per se, but to assess the potential outcome of the situation, whether it is positive or negative. The term Sunnafrank (1986) uses for this is "predicted outcome value theory" (p. 3).

Uncertainty reduction theory in communication relationships involves the two psychological components of self-awareness and knowledge of others. In 'objective self-awareness' the person focuses on the self rather than the objects of the environment. In 'subjective self-awareness' the self is viewed as peripheral within a continual stream of experiences (Littlejohn, 2002, p. 243). Any 'self and other' regarding behavior, such as those behavioral patterns frequently involved in the interpersonal aspects of business communication, involve the concept of understanding or being aware of the predictive uncertainty of others as a measure of expectations, and may involve a notion of explanatory uncertainty in comprehending their actions (Littlejohn, 2002, p. 245). However, ambiguity is not only expressed orally and behaviorally. Causes of ambiguity in written business communication include "ignoring the reader or audience, a lack of professional pride, a lack of confidence, inexperience, and writing for the wrong reasons, and strict requirements" (Piotrowski, 2005, pp. 1-2). Some ambiguity is clearly positive in so far as it is motivational, however, low ambiguity in the group context may result in the negative state of 'groupthink' which is best understood as a "deterioration of mental efficiency, reality testing, and moral judgment that results from in-group pressure" (Verderber, Verderber, & Berryman-Fink, 2008, p. 234). Groupthink is all the more pernicious because group members may not be aware of its onset or consequences. Milgram's (1963, 1974) experiments on obedience to authority figures at Yale University are a classic case in point.

The Dependence on Language in the Professional Environment

Any professional workplace context is dependent on language use. While different professions (such as law, medicine or accounting) may have specialized terminologies, many professional workplaces get by with the efficient use of Plain English. Having determined that ambiguity is a potential cause of inefficiency in the workplace, it is now possible to identify three causes of ambiguity: a) semantic ambiguity – words may have more than one meaning, requiring discernment from the listener, b) people have difficulties in the attempt to explain meaning, and c) people of different languages or cultural groups may have inter-lingual difficulty in understanding one another (Holland & Webb, 2006). A factor in remediation involves simply taking the extra time needed to know and understand the other person and comprehend their point of view. Frequently trying to clearly separate the message from the messenger is also beneficial, but not wholly so.

Ambiguity in communication has benefits and risks. While generally scientists and business people seek certainty, clarity, and non-divergence of meaning, in other contexts what is required is the deliberate preservation of uncertainty, requiring ambiguity. This may arise in situations where the facts are unknown or in which the event may take place sometime in the future. Ambiguity is thus used to keep open future possibilities. As Komesaroff (2005) states, the "construction of the future from within the present as an open array of possibilities … is what we call 'hope'" (p. 632). Hope is dependent on ambiguity. While hope certainly plays a role in any future-oriented business, business communication can be complex, involving a heterogeneous mix of discursive practices which arise from different discourses. These may include: scientific knowledge, economics, social insight, marketing, and philosophical reflection. Clearly, crystal-ball-gazing is not simply a fairground attraction.

Tolerance for ambiguity is also related to group motivations. Salazar (1996) states that, "If individuals composing the group have little tolerance for ambiguous situations, they may be perceived as being too complex" (p. 177). However, individual characteristics and tolerances toward ambiguity may differ from group uses or from strategic use of ambiguity. Menz (1999) argues that a "continuous switching between routine and non-routine helps to create the delicate balance between ambiguity reduction and ambiguity maintenance, which is essential to the continued existence of organizations" (p. 104).

Reverting to clear and simple communication may be the default position in business practices; it is not always possible to traverse all communication divides. The ability to effectively use ambiguity in business communication is an everyday competence. Speech within the business environment is often not a solitary or impersonal activity but rather a process of shared discovery and of contested meanings. It is useful to bear in mind that there is "no exact or complete transmission that remains unchanged within any cultural system of meaning" (Komesaroff, 2005, p. 633).

Ambiguity in Job-Related Messages

Discounting a normal range of tolerance for communication ambiguity, Krayer and Bacon infer that differences in the perception of ambiguity in a message is "significantly related to differences in role ambiguity perceived by the worker on the job" (p. 90). As role ambiguity may cause stress and organizational dysfunction, the knowledge that role ambiguity and communication ambiguity are correlated may mean that a reduction in the former may result in a reduction in the latter. Thus the ambiguity of job-related messages may reduce in an organization in which roles are clearly defined.

Pickert (1981, p. 56) has described how the ability to provide a solution for an ambiguous task may enact 'decentration' or 'reversability,' (the ability to focus on two or more aspects of a problem simultaneously and relate them together) in two ways. Firstly, to interpret and respond to an ambiguous communication may involve the consideration of another person's point of view (involving concrete operational thought). As the respondent may realize that they lack information, they will then ask questions. Social decentration may follow as the logic of the problem presented is considered but also the responses of other interpretants (Pickert, 1981).

Furthermore, cognitive dissonance may be the result of communication in the business organization as the result of unexpected appropriation or blame or simply changing market-circumstances. Dissonance occurs as the result of an unexpected correlation between concepts or events which may produce tension or stress as the result of a perceived need to change (for example, either attitudes or behaviors). The overriding consequence of cognitive dissonance is the desire to reduce it; however, in the extreme this may produce "situational avoidance" (Littlejohn, 2002, pp. 126-127).

Means (2010, p. 45) and Piotrowski (2005) outline 10 principles that provide a rule of thumb for managing ambiguity in the business environment:

1. Organize your thoughts
2. Maintain a professional attitude
3. Refrain from making judgments about others
4. Keep an open mind
5. Don't make assumptions or jump to conclusions
6. Keep emotions in check
7. Be slow to take offence
8. Give others the benefit of the doubt
9. Keep control of yourself
10. Ask what your immediate purpose is, what is the scope of the problem, what are the constraints or your response?

Following these 10 principles may alleviate some, but probably not all of the ambiguity associated with job-related messages.

Strategic Ambiguity Within the Business Organization

Although some of our experiences with ambiguity may teach us that while being clear in communication is a worthy goal, there exists a body of thought which argues that clarity in communication is 'non-normative' and not a sensible standard against which to assess individual or organizational effectiveness. Rather, ambiguity can be seen as a continuum between exact understanding and nonsense. We are more likely to achieve rapport with someone who communicates such that we can understand them and others are more likely to achieve rapport with us if we can communicate so as to be understood. For dependable patterns of action and independent initiatives to coexist in the business environment, a consensus of meaning is sufficient but not necessary. However, as Salazaar (1996) states that ambiguity may have an effect on group strategy for decision-making because " ... under conditions of uncertainty and ambiguity, individual decision makers engage in search processes to better define the problem confronting them .... [S]earch involves assessing information regarding alternatives, establishing criteria, and examining outcomes and payoffs associated with those outcomes" (p. 176). The process of searching can itself be creative but risks divergence unless clear parameters are set. A further response of strategic ambiguity is convergence. This occurs when people may share "fantasy themes or a rhetorical vision" which gives them a sense of identification with a shared reality (Littlejohn, 2002, p. 158). Convergence may occur informally within an organization and only tangentially be related to it, or more formally through public relations campaigns.

In order to be able to adapt, Menz (1999) argues that an organization needs to be able to effect "a continuous balance between adaptation and adaptability, between stability and flexibility" (p. 105). Furthermore, Eisenberg (1984) has argued that strategic ambiguity is essential to a business organization insofar as it, 1) promotes unified diversity, for example, by contributing to group identity formation, 2) facilitates organizational change, by signaling aspirational values, and/or, 3) amplifies existing source attributions and preserves privileged positions by "mythologizing leadership roles" or by minimizing commitments that are difficult to break (p. 227).

Ambiguity and Organizational Climate

The positive view of some levels of strategic ambiguity within organizations is partly a response to internal recognitions and partly a result of the desire to achieve external objectives. For example, employees might be enthused by a successful advertising campaign depicting their business or embarrassed by inadequate representation.

Moreover, they may become frustrated by the inability to effect changes in company public relations and become cynical.

The role of the recognition of value of ascribing 'cognition' to the worker in an organization in historical terms is a relatively modern concept. Most efficient and productive workplaces today will value some degree of workplace autonomy within organizational structures. While the desire for accountability will remain the predominant factor, as Eisenberg (1984) suggests, there is a shift in most progressive workplaces to seeing organizational participants as "thinking individuals with identifiable goals" (p. 227). In order to distinguish ambiguous communication firstly from the interference and noise and from the disruption to channels of communication, it must be understood that "complex systems are not dominated by linear and causal relations, but by non-linear ones" (Menz, 1999, p. 105). The difficulty here is that non-linear relations may be a cause, by-product, or indeed causally irrelevant to the creation of ambiguity, but *are* seen as relevant to its perception. In ambiguous workplace situations, it is difficult to distinguish between cause and effect because frequently communication may be attributable to the organizational 'climate' rather than to a defined source.

Within many organizations, the role of manager is seen to be taken by people who are skilled symbolic communicators. Often managers use language effectively because it is a by-product of political, dramaturgical, and language skills necessary for organizing resources and people. Competent communicators may use strategic symbols to accomplish their goals, but the communicative change of this language use may not always be open or clear (Eisenberg, 1984, p. 227).

Furthermore, as objectives change, the goals of communicators may also change. Meaning is rarely unitary or consistent; often people in the workplace have multiple and sometimes conflicting goals which they orient towards. As people in organizations confront multiple situational requirements, they may respond to communicative strategies which do not always minimize ambiguity. As Eisenberg (1984) comments, the climate of a business may also shift from an ideological adherence to clarity to one of a contingent strategic orientation; this may or may not be communicated among the various levels of the organization (p. 228). In any organization there may be three or more levels of communication reflecting different uses and tolerances of ambiguity. At lower operative levels, there is often tighter control over ambiguity than at middle-management level, whereas higher levels of management may use both ambiguous and non-ambiguous language in everyday language use.

Eisenberg (1984) compounds the complexity of the use of strategic ambiguity in organizational communication by suggesting that while explicit communication may be a concept of cultural assumptions, it is "not a linguistic imperative" (p. 228). People will vary their language along a continuum of explicitness and indirection, depending on how they may read another person's desires, aims, or understandings (Eisenberg, 1984, p. 228). Ambiguity may thus be seen as a resource of language which may intentionally or otherwise be employed to strike a balance between being understood, "not offending others, and maintaining a self-image" (Eisenberg, 1984, p. 228). Furthermore as Salazar (1996) suggests, in decision-making situations characterized by minimal ambiguity in which "members are homogeneous" with regard to information and the "task is not novel," communication is likely to play a minimal role (p. 179). This may be compared with organizational situations of high ambiguity in which the motivation to "persuade, exchange information, check for errors in reasoning" – all necessary factors for effective decision-making – will be stronger (Salazar, 1996, pp. 178-179).

Frequently any communicated message will deviate from a hypothetical ideal. Ambiguity may be defined in terms of message attributes (lack of specificity, abstract language, absence of a course of action) and receiver interpretation (perceived equivocality of the message) (Eisenberg, 1984, p. 229). Eisenberg (1984) argues that the concept of an ideally clear message is misleading in so far as clarity is a relational variable a continuum which "reflects the degree to which a source has narrowed the possible interpretations of a message and succeeded in achieving a correspondence between intentions and interpretation of receiver" (p. 230).

In a turbulent business environment where there are many impacting external or internal variables, ambiguous communication is a rational method used by communicators to orient towards multiple goals (Eisenberg, 1984, p. 239). In any organization there are conflicts between centralization and decentralization, the individual and community, self-determination and security. As such, strategic ambiguity may promote unified diversity. The issue of divergent goals may be manageable, not necessarily by consensus, but through the "development of strategies to preserve and manage differences" (Eisenberg, 1984, p. 231). A variation of strategic ambiguity is equivocal communication. This is described as non-straightforward communication which may be ambiguous, tangential, or evasive. Equivocal communication is usually indicative of avoidance and is used when all other forms of communication might lead to negativity (Tyler, 1997, pp. 59-60).

In complex business organizations with many members and multiple organizational values and interpretations, a sense of unity may by derived from points of symbolic divergence. Ambiguity may be used strategically to encourage creativity and guide against the acceptance of "one standard way of viewing organizational reality" (Eisenberg, 1984, p. 231). Sometimes the role of leaders is to provide meaning for followers. The language required may be abstract, evangelical, and even poetic.

Interpreting Meaning

Thus, there are many business contexts in which a literal meaning cannot be taken, or an intended meaning may not be understood. For example, if you are discussing your work allocations with your manager and she or he tells you that you have "five-floating," it could mean he or she wants to do the 'high-five' with you, or it may mean that you have five days of leave in lieu. Asking questions and seeking clarity can help but symbolic language may also transcend the context it is spoken in.

A further ethical dilemma with strategic ambiguity in business is that it may or may not involve a mischief element. Was the statement intended to be ambiguous and therefore to mystify? Can it be discounted as 'only a symbol' and hence meaningless? From the point of view of interpretation, *there could be fish hooks with that*, but ethically, clear commentary is preferable in a business setting when dealing with day-to-day issues.

The golden rule of interpreting words in context and giving them their ordinary meaning (unless this produces an absurdity) may work in formal situations of business activity, but not always in less formal situations due to the transitory and oral nature of business communication. The allocation of correct 'weighting' to the communication context in the workplace is crucially important. This may be derived from both fields of experience and organizational knowledge.

Ethical Uses of Strategic Ambiguity

As Eisenberg (1984) suggests, the uses of ambiguity within organizations are multiple but coalesce around issues relating to group leadership. If used creatively and relatively benignly, strategic ambiguity can achieve organizational goals in an ethical manner. It can be used to provide inspiration that will imprecisely guide a group towards a desired aspiration, lead organizational change, or motivate productive organizational behavior. Further identifiable benefits of strategic ambiguity include: holding strained relations together, allowing a group to employ a single voice, facilitating change, allowing for adaptation, creating durable meanings, maintaining standing, and 'character insurance,' avoiding costly commitments, and preserving future options or courses of action (Eisenberg, 1984).

Moreover, at an interpersonal level, strategic ambiguity can, a) facilitate relational development, b) control what people share of private opinions, beliefs, and thus avoid conflict, c) act as a buffer of deniability (essential to the moderation of different views), and d) provide a useful compromise between known and unknown. However, it is necessary to stress that a distinction needs to be made between strategic ambiguity used at an interpersonal level and taking 'strategically ambiguous' communication personally. The latter may be ethically inadvisable in circumstances in which relationship needs conflict with organizational expectations or protocols.

Furthermore, there is a fine line to be drawn between the use of strategic ambiguity in public relations campaigns, and deception. Here it is relevant to make a distinction between the use of strategic ambiguity in public relations campaigns that reflect realities of the business' operating environment and those which seek to conceal or present alternative realities. There are no uniformly established ethical guidelines but today's society of social media prolificacy is finely attuned and potentially rapidly mobilized to expose major discontinuities in issues of public interest.

Krohn (1994) has argued that the Sapir-Whorf-Korzybski hypothesis may be applied in the context of training business communication students in ethical practice by discouraging words that denote violence and including general semantics training. Mahin's (1998) view is that "...the very act of communicating in the social context of a business culture implies an ethical basis, a respect for persons" (p. 74). However, the fact that there is a lack of a common framework for deciding what is ethical language use in business practice tends to detract from the concept that openness in communication is a factor which affects business performance (Nelson, 2003).

Ambiguity and Crisis Management

As Ulmer and Sellnow (2000) have suggested, crises in which there is a high degree of organizational ambiguity can result in benefit to the organization if handled effectively. In crisis situations the idea is to "reduce communication ambiguity" (Ulmer & Sellnow, 2000, p. 143). One of the difficulties in reducing ambiguity in crises is that businesses need to maintain the support of many groups, including customers, employees, stockholders and regulatory agencies. Efficient communication is therefore essential, since reducing contradictory information lessens the ambiguity of crises and allows for future planning. Here reductionism may be useful (reducing a problem to its smallest parameters). In this respect, ambiguous communication may allow divergent interpretations to coexist and diverse groups to work together (Weick, 1988, p. 305). However as Ulmer and Sellnow (2000, p. 147) suggest, strategic ambiguity may be ethical when it involves conveying complete and unbiased information but unethical when biases or gaps in knowledge are evident.

Ultimately the discourse of business communication may be accepted or rejected on the grounds of reasonableness. Thus understanding the ethical complexities of losing balance in the communication with relevant stakeholders is important in understanding and rectifying communicative practices to resolve ethical tensions in a crisis. Cheney (1991) observes that understanding ambiguity may lead to a decentring of the self within the organization by making the worker view themselves as a subject within a system they do not fully understand. In normal organizational communication, not all puzzles have to be solved straight away.

Ambiguity and Risk

Ambiguity can be understood as being similar to economic risk, a term used to describe a situation in which an investment is made but of which the outcome is uncertain. In some forms of risk, outcomes are uncertain but probabilities are known. In others, termed economically ambiguous, the probabilities of uncertain outcomes are unknown (O'Neill & Kobayashi, 2009, p. 645). People prefer to take 'risky' decisions rather than ambiguous decisions, which suggests that people are averse to making decisions in "low information environments" (O'Neill & Kobayashi, 2009, p. 645). Furthermore, neuroeconomic studies by Ellsburg (1961) and Hsu, Bhatt, Adolphs, Tranel and Camerer (2005) have discovered evidence for brain regions preferentially activated by ambiguity (the *frontal cortex* and *amygdale*) and risk (the *parietal cortex* and *striatum*), implying that they may be separately encoded in the brain. So there is physiological evidence that the human brain processes ambiguity and risk in different ways.

Ambiguity in Acquisitions and Mergers

Risberg's study of ambiguity and communication in cross-cultural business acquisitions reveals that the process of company acquisitions and mergers is for the acquired company, a threatening experience. One reason is that the present and future frequently become ambiguous. Problems are seen to develop from lack of sufficient information or a function of different values (Risberg, 1997). Therefore consistency of information between acquiring and acquired companies is important in creating meaning out of uncertainty. Differences in organizational cultures can be due to differences in ethnicity, gender, nationality or ideology or subcultures with conflicting assumptions (Risberg, 1997). Risberg (1997) suggests three perspectives that may be used to study corporate culture: the integrative perspective, the differential perspective, and the ambiguity perspective" (p. 258). The integrative perspective emphasizes commonalities between organizational cultures, while the differential perspective stresses inconsistency and lack of consensus. The business culture is seen as either harmonious or conflicting. The ambiguity perspective combines aspects of each.

Many of the failures in acquisitions and mergers are due to the ambiguities produced from cultural clashes. These occur when companies refuse to find commonalities and instead see differences. Unlike the integrative perspective, the differentiation perspective does not deny ambiguity; there may be different sub-cultures within any organization. Multiple perspectives in any one organization and a climate neither wholly harmonious nor conflicting are more likely to be formed in an organizational culture viewed from an ambiguity perspective.

Ambiguity and Leadership

While most workplaces would have ideals of security, community, respect, authority, and clarity in communication, in the era of social media, the worldwide web, globalization, and global warming, the pace of change within organizations has increased dramatically. As a consequence, many organizations need to remain agile and responsive to changing internal and external environments. However there is a fine line between the communication of organizational values and the communication of organizational change in which planning sometimes gives way to adaptation. As Amorium (2010) suggests, business organizations need to know who they are (how they are defined in their operations) and what they want (or want to become), because the "how of their plans will be a moving target" (p. 2). Clearly, visionary leadership and the ability to implement plans are on a continuum with variable points of intersection. Consequently, in times of organizational change, dealing with ambiguity is a leadership skill. Hooper (2007, p. 1) defines the following characteristics of leadership skills:

Leaders:
• Can effectively cope with change
• Can decide and act without having a total picture
• Aren't upset when things are up in the air
• Can comfortably handle risk and uncertainty
• Are future-orientated
• Can handle volatility, uncertainty, and complexity
• Identify threats and opportunities in business practice.

Thus leadership skills for 'uncertainty' are desirable and may in fact be necessary qualities for the modern workplace. Leaders often attract praise or blame for their judgment, but suspension of judgment may in fact be more valuable in many situations in which ambiguity results from excessive uncertainty or change. On the other hand, clarity of communication may enhance or inhibit workplace anxiety. Mintzberg (2009) critiques a growing trend he terms 'macroleading', characterized by "leaders who manage by remote control, disconnected from anything except the big picture" (p. 9). However, the opposite trend of 'micro-managing' may also inhibit organizational direction and growth by dampening down necessary sparks of creativity.

For any leader in today's business market, opportunities and challenges must be appreciated from multiple viewpoints and not just seen as traits to be expressed from a personal leadership bias. Causes of conflict within an organization are often difficult to diagnose from a single perspective. Thus a leader's role must sometimes be to set reasonable goals and to 'disambiguate' them. However, providing a clear direction, synchronizing the motivations of others, and communicating adjustments within an organization may involve both clear instruction and ambiguation. Other qualities of a leader's toleration to ambiguity include listening well, thinking divergently, and the setting up of incremental dividends to reward the efforts of workers.

Anthony (2010, para 4) argues that complexity, sudden shifts in the basis of competition, and global competitors are the "new norms" of constant change which face tomorrow's global leaders. While giving people more responsibility helps them refine skills, the acquisition of new skills as the result of ambiguous threats and challenges may also be a driver of business capability. Thus Anthony (2010) suggests rather than a scale being a measure of success, giving leaders "smaller ambiguous challenges" may instead result in the acquisition of competencies necessary for climbing the corporate ladder (para 7).

Conclusion

There are many effective business writing books which will tell the student how to structure a written report and lessen ambiguity in the discursive business environment, (for example see Piotrowski's *Effective Business Writing*), but these frequently do not provide a full account of the characteristics and utility of ambiguity in business communication. Ambiguity in business and organizational communication needs to be understood more thoroughly as the inevitable result of communication in high and low context workplace protocols, as the changeability of meaning as the result of language use in interpersonal and strategic communication contexts, and as a consequence of the uncertainties of change management in organizations.

Managers and staff may need to be trained to be able to use conflict management strategies effectively and deal with situations of ambiguity and uncertainty. This may involve deeper understanding of aspects of interpersonal communication or the uses of strategic ambiguity within an organization. As Robbins (1993) points out, conflict may result from incompatibility over goals, differing interpretations of facts, disagreements about behavioral expectations, and from arguments over resources. However, as Sayers (2005) suggests, sometimes conflict "cannot and should not be resolved" (p. 83).

While one strategy for dealing with ambiguity is simply to learn to tolerate it, by gradual frequency and acclimatization to exposure from the people who employ it, a creative element in strategic communication can be found which can work to an organization's advantage. To use strategic ambiguity takes confidence because it also contains a risk of indirect communication or that the audience simply will not understand it.

Many working environments may go through periods of structured antagonism in which general staff values of freedom may conflict with values of managerial control. Conflict may arise from semantic difficulties, misunderstandings, lack of information, or information distortion. Skills that managers of ambiguity in business organizations require are being able to reconsider boundaries of people's positions, thinking creatively to find new solutions, emphasizing relatedness rather than polarizing views, the willingness to work through any problem, and not necessarily to seek to end conflict but to "manage it properly" (Sayers, 2005, p. 93). While there may be an important practical consideration in maintaining clear and unambiguous communication in everyday procedural matters within organizations, when managed carefully, both tactical and strategic ambiguity can add value to an organization's communication if used in an ethical manner. Finally, the best practical advice is that when language use seems interpersonally ambiguous, it is far better to interpret it positively than negatively.

Acknowledgements: My thanks are due to Dr Elspeth Tilley and *Prism* for comments and permissions.

References

Amorium, M. (2010). *Ambiguity and the new business normal.* Retrieved from: http://www.maurilioamorium.com/2010/12/ambiguity-and-the-new-business-normal/

Anthony, S. (2010). 'Grooming leaders to handle ambiguity'. *Harvard Business Review.* Retrieved from: http://blogs.hbr.org/Anthony/2010/07/groomin g_leaders_to_handle_ambiguity.html#

Berger, C. (2003). Uncertainty reduction theory. In E. Griffin (Ed.)., *A first look at communication theory.* (pp. 142-152) New York: McGraw Hill.

Bernoulli, D. (1954/1738). 'Exposition of a new theory on the measurement of risk', (L. Sommer, Trans.). *Papers of the Imperial Academy of Science of Saint Peterburg*, *5*, 175–192.

Bradley, A. (2008). *Derrida's of grammatology: An Edinburgh philosophical guide*. Edinburgh: Edinburgh University Press.

Buller, D., & Burgoon, J., (2003). Interpersonal deception theory. In E. Griffin, *A first look at communication theory*. (pp. 95-110). New York: McGraw Hill.

Cheney, G. (1991). *Rhetoric in an organizational society*. Columbia, SC: University of South Carolina Press.

Cheney, G., Chistensen, L., Thoger, Z, Ganesh, T., Shiv, G. (2004). *Organizational communication in an age of globalization: issues, reflections, practices*. Prospect Heights, Ill: Waveland Press.

DeVito, J., O'Rourke, S., O'Neill, L. (2000). *Human communication*. Auckland, New Zealand: Pearson Education NZ Ltd.

Dwyer, J. (2009). (4th ed.) *Communication in business: strategies and skills*. Frenchs Forest, NSW: Pearson Education Australia.

Eisenberg, E. (1984). 'Ambiguity as strategy in organizational communication'. *Communication Monographs*, *51*, 227–242.

Ellsberg, D. (1961). 'Risk, ambiguity and savage axioms'. *Quarterly Journal of Economics*, *75*, 643–679.

Empson, W. (1996). *7 types of ambiguity*. New York: New Directions.

Holland, J., & Webb, J. (2006). *Learning legal rules: A students' guide to legal method and reasoning*. New York: Oxford University Press.

Hooper, D. (2007). Dealing with ambiguity – part 1. Retrieved from: http://www.buildingfutureleaders.com/uploads/4/1/1/4/411493/microsoft_word_-_dealing_with_ambiguity--part_1.pdf

Hsu, M., Bhatt, M., Adolphs, R., Tranel, D., & Camerer, C. F. (2005). 'Neural systems responding to degrees of uncertainty in human decision-making'. *Science, 310*, 1680–1683.

Kahn, R. L., Wolfe, D.M., Quinn, R. and Snoek, J. D. (1964). *Organizational stress: Studies in role conflict and ambiguity*. John Wiley & Sons, New York.

Kahneman, D. & Tversky, A. (1979). 'Prospect theory: An analysis of decision under risk'. *Econometrica, 47*(2), 263–291.

Komesaroff, P. A. (2005). 'Uses and misuses of ambiguity: Uses of ambiguity'. *Internal Medicine Journal, 35*, 632-633.

Krayer, K., & Bacon, C. (1984). 'Communication ambiguity in job-related messages'. *Communication Research Reports, 1*(1), 88–90.

Krohn, F. (1994). 'Improving business ethics with the Sapir-Whorf-Korzybski hypothesis in business communication'. *Journal of Education for Business*, *69*(6), 354–359.

Littlejohn, S. (2002). *Theories of human communication*. Belmont, CA, USA: Wadsworth/Thomson Learning.

Mahin, L. (1998). 'Critical thinking and business ethics'. *Business Communication Quarterly*, *61*(3), 74-78.

Mead, G. (2003). Symbolic Interactionism. In E. Griffin, *A first look at communication theory*. (pp. 55-65). New York: McGraw Hill.

Means, T. (2010). *Business education* (2nd ed.). Mason, USA: South Western, Cengage Learning.

Menz, F. (1999). 'Who am I gonna do this with? Self-organization, ambiguity and decision-making in a business enterprise'. *Discourse and Society*, *10*(1), 101-128.

Milgram, S. (1963). 'Behavioral study of obedience'. *Journal of Abnormal and Social Psychology*, *67*(4), 371–378.

Milgram, S. (1974), *Obedience to authority: An experimental view*. New York: Harper & Row.

Mintzberg, H. (2009). *Managing*. San Francisco: Berrett-Koehler Publishers.

Nelson, R. (2003). 'Ethics and social issues in business: An updated communication perspective'. *Competitiveness Review: An International Business Journal incorporating Journal of Global Competitiveness*, *13*(1), 66 -74.

O'Neill, M., & Kobayashi, S. (2009). 'Risky business: Disambiguating ambiguity-related responses in the brain'. *Journal of Neurophysiology*, *102*(2), 645-647.

Pickert, S. M. (1981). 'Ambiguity in referential communication tasks: The influence of logical and social factors'. *The Journal of Psychology*, *109*, 51-57.

Piotrowski, M. (2005). *Effective business writing*. New York: Collins.

Risberg, A. (1997). 'Ambiguity and communication in cross-cultural acquisitions: towards a conceptual framework'. *Leadership & Organization Development Journal*. *18*(5), 257–266.

Robbins, S. (1993). *Organizational Behaviour*. Englewood Cliffs, N J: Prentice Hall.

Salazar, A. J. (1996). 'Ambiguity and communication effects on small group decision-making performance'. *Human Communication Research*, *23*(2), 155-192.

Sayers, J. (2005). Managing conflict at work. In F. Sligo & R. Bathurst (Eds.). *Communication in the New Zealand workplace: Theory and practice*. (pp. 83-94). Wellington: Software Technology New Zealand Ltd.

Sunnafrank, M. (1986). 'Predicted outcome value during initial interactions'. *Human Communication Research*, *13*, 3-33.

Tyler, L. (1997). 'Liability means never being able to say you're sorry: Corporate guilt, legal constraints and defensiveness in corporate communication'. *Management Communication Quarterly*, *11*(1), 51-73.

Ulmer, R. & Sellnow, T. (2000). 'Consistent questions of ambiguity in organizational crisis communication: Jack in the box as a case study'. *Journal of Business Ethics*, *25*, 143-155.

Verderber, R., Verderber, K., & Berryman-Fink, C. (2008). *Communicate!* (12th ed.). Belmont, CA: Thomson Wadsworth.

Wallerstein, I. (2004). *The uncertainties of knowledge*. Philadelphia: Temple University Press.

Watzlawick, P., & Weakland, J. (1977). *The interactional view*. NY: WW Norton.

Weick, K. E. (1988). 'Enacted sensemaking in crisis situations'. *Journal of Management Studies*, *25*(4), 305-317.

9

ORGANIZATIONAL COMMUNICATION
Management, Leadership and Self-Esteem

As Corrigan (1999) suggests, the primary experience of management is "responsibility without control" (p. 14). This experience is made more complex by the fact that management takes place along a spectrum of both analytic and affective dimensions, being the product and stimulation for the intended work activity. In many companies, the creation of a successful business culture is seen as one of the determinants of creating the necessary conditions for leadership growth. A successful business culture has two purposes: it ensures that the culture of the organizations needs to be taught to its leaders, and it indicates that the organization is creating the necessary conditions for leadership to flourish. Many business environments today need to adapt to unpredictable pace and change. The primary aim of this chapter is to examine the main tenets of effective management – to show the characteristics of management models which work well and those that do not. The secondary purpose is to delineate the fundamental tenets of sound leadership practices. The third aim is to show how both management and leadership are integrally related to self-esteem.

Uncertainty necessitates stronger leadership in organizations operating in current climates and market conditions. If internal and external conditions are known, then foresight is not required. However, leaders are needed when it is difficult to see into the future. Such foresight is a complex requirement, and some leaders may become immobilized by not knowing what is going on in their business environment or by being required to operate 'beyond what they can see directly in front of them.'

Corrigan (1999, p. 16) suggests that in the current business environment, which is influenced by the global economy and technological change, quality leadership is needed all the more as it gives the organization direction in an environment of rapid change. The basic task of strategic leadership is to identify threats and opportunities. However, as Gavetti (2011) states, "strategic leaders are not omniscient" (p. 120). This has two consequences: first, it may mean that advantages and opportunities are available but remain undetected. Secondly, because organizational strategists have similar mental representations, they may overlook similar opportunities. Gavetti (2011) terms these as cognitively distant opportunities that may require changes in an organization's identity. Consequently, leaders frequently need to be able to both develop strategic ability and to gauge the depth of field of a strategic issue.

First, a major task of managers is to take responsibility for those under them (Corrigan, 1999, p. 16-17). The cognitive dissonance differential of the learning environment (the gap between a current and future state of knowledge) is also replicated in management pedagogy. As Corrigan (1999) suggests, managers are seldom satisfied with the status quo and "restlessness is at the core of management" (p. 17). However, concomitant with the need to overcome inertia is both an awareness of the subjective and objective factors influencing the environment, and a disinterest in "authority for itself" (Corrigan, 1999, p. 21).

Management also depends on effective communication, although the personality of the manager most often influences the way their management regime is enacted. Secondly, managers need to be mindful that the operating environment may often be neutral, that is, people bring emotions to their workplaces but the structural environment itself has no emotional value. While managers may communicate with their staff with a range of techniques, including both literal and figurative language, a fundamental learning goal is that they take responsibility without the "*control* [emphasis added] that is necessary to *guarantee* [emphasis added] success" (Corrigan, 1999, p. 25).

One of the most unsettling experiences for managers is that they may take responsibility for matters beyond their control. Corrigan (1999, p. 26) identifies three main different kinds of manager: those who find that their lack of control makes responsibility difficult;

those who act as if they can control everything but cannot; and those who live with the experience of not being able to control everything they need to, but take responsibility for the issues their job describes. Thus the main task for many managers is to learn how to live with anxiety, to cope with it, and to continue to perform. Some bad managers look for issues that they should be responsible for and become frustrated at their lack of ability to control everything, thus leading them to deny any responsibility, causing an inability to act (Corrigan, 1999, p. 26).

The relationship of managers to authority is complex. On the one hand, it may be ascribed by a higher authority and, on the other, it follows from the natural performance of work-related objectives. Most modern organizations believe that authority is achieved rather than ascribed. Managers gain authority by allegiance with staff following from their performance of the job – respect is thus earned through action (Corrigan, 1999, p. 37). However, managers may acquire their right to action from both title and action. Those who work more closely with staff to achieve actions may gain more authority from allegiance than those who achieve power through their actions alone.

Corrigan (1999) suggests that the second main obstacle for leaders is losing touch with reality. Consequently the idea that it is good to work with other people is rooted in the necessity of "authority to involve itself with followers" (Corrigan, 1999, p. 39). Leaders and managers thus need to guard against losing touch with reality. Part of this may come from awareness of the ways in which we may affect the world and are affected by it. As Corrigan (1999) suggests, "We are not determined by the world around us, rather we act upon it and intervene. We are however affected a great deal by the social, economic, and psychological conditions of our time" (p. 41). Similarly, if managers become disconnected from their staff, their leadership may become more fragile, "where power resides in ... a one-dimensional authority, the world can fall apart in chaos" (Corrigan, 1999, p. 59). A third common mistake of managers is defending the organization from change. Building rigidity into managerial actions may make real change problematical. The overriding aim of the manager is to make things better by acting incrementally to bring about positive change (Corrigan, 1999, p. 66). Fourthly, the personalization and abstraction of power from the organization by managers results in leadership fragility, for if power resides in an individual then it may dispensed on a casual basis. Corrigan (1999, p. 79) suggests that people who view power in this way may be jealous of others encroaching on it or be nervous that some of their power can simply be taken away. A fifth obstacle for good management is that in situations where managers are frightened of the talent beneath them, they may reward mediocrity (Corrigan, 1999, p. 79).

This concept is related to office politics, the pursuit of power games within the organization without clear direction or for self-interested reasons (Corrigan, 1999, p. 85). As Corrigan (1999, p. 87) suggests, the three main motivations within the organization include: personal recognition, salary gains, or the power to push people around. However positive motivations for organizational advancement include ambition to achieve higher ideals, to make the organization more morally upright, or to stop the advancement of dubious motives, positions, policies or actions.

The motivation of managers is thus a profound contributor to organizational health. Inhibitors of organizational health may include the psychological need to manipulate the world to fit the individual's desires. Similarly, powerful ambition, if unchecked, can ruin lives and organizational health. Individual ambition can be contained within an apparent loyalty to a wider group. There is thus a general perception that management is difficult because it has some unpleasant responsibilities, such as a readiness to discipline or fire staff. Sometimes managers may appear manipulative because they may need to consider people, events and organizations as objectives or in a utilitarian manner to achieve their aims.

Raynor (2009, p. 436) argues that educational leadership involves making linkages between theory (knowledge) and provision (practice), which combine to produce an applied theory (praxis). Consequently, as well as providing service to the organization, producing new knowledge may also contribute to the manager's phronesis (practical thought) and act as a catalyst for the production of learning as a purpose and outcome (telos), as well as form the basis of a leader's professionalism. Furthermore, Raynor (2009) suggests that leadership cognitions involve "sense-making, problem-posing, decision-making and critical reflection" (p. 435).

Corrigan (1999, pp. 72-73) identifies four main variables in managerial leadership: 1) the characteristics of the leader, 2) the attitudes, needs and characteristics of the staff, 3) the characteristics of the organization, and 4) the wider economic, social, and political climate. That is, the leader thinks and acts in a particular time, place, or structure. Consequently, it is more sensible to view leadership as relational between staff, manager, and organization than to see it as an innate characteristic of certain individuals (Corrigan, 1999, p. 72-73).

Uncertainty Toleration in Leadership

Hodgson and White (2001, pp. 98-99) identified 16 clusters of behavior that individuals either use to manage ambiguity and cope with uncertainty or which act as restrainers – so-called 'anti-skills' that hinder the ability to tolerate and act under conditions of ambiguity.

Table 1. Characteristics of 'enablers' and 'restrainers' in leadership tasks

Enablers	Restrainers
Mystery seekers – high curiosity, fascinated by what they don't know.	**Poor transitioners** – have difficulty shifting from one task to another and a limited repertoire.
Risk tolerators – an ability to make choices with incomplete information/see mistakes as a way to learn.	**Wet blankets** – dampen the energy of the organization.
Future scanners – want to understand how a business operates and consider its future states/curious about what the future may bring.	**Conflict avoiders** – overly accommodating to others and highly averse to interpersonally heated situations.
Tenacious challengers – tirelessly solve problems, seek solutions.	**Muddy thinkers** – exhibit self-inflicted confusion.
Exciters – want everyone to be energized by what they do/strive to make work fun for others.	**Complex communicators** – use overly complex language.
Flexible adjusters – exhibit two tendencies: an ability to admit they are wrong; or an ability to sell change to people against change.	**Detail junkies** – obsess over smaller tactical issues to the exclusion of strategic ends.
Simplifiers – able to take complicated ideas and simplify them.	**Narrow thinkers** – exhibit tunnel vision, focus on the moment and are blind to new possibilities.
Focusers – identify and attack critical actions.	**Repeaters** – tethered to the past and continue to rely on actions which may no longer be relevant.

White and Shullman (2010, p. 94) argue that leadership development has taken us beyond the concept that leaders are born, and that organizational environments are universally stable. Instead they have suggested that flexibility is inherently necessary in any dynamic organizational environment. The dynamic and flexible quality of the business environment may be due to organizational evolution, the transition from command and control leadership to participatory paradigms, organizational learning, or an ability to navigate ambiguity (White and Shullman, 2010, p. 94).

Consequently, being an effective leader involves an ability to operate under conditions of uncertainty and to deal with ambiguity. Those who have an ability to operate under conditions of ambiguity are better able to keep their options open. In many ways, leadership is a "mutual agreement between leaders and followers" (White and Shullman, 2010, p. 95). Participation, empowerment of others, and technical expertise are all factors of effective leadership. However, the model of effective relationship management is not in itself sufficient; rather an ability to adapt to novel problems is increasingly paramount.

Leaders also need to know when to draw on the collective wisdom of the organization. Under these conditions, the ethical use of strategic ambiguity can be an enabler. White and Shullman (2010) suggest that effective leaders are those who have an ability to accomplish tasks with an ability to identify those who can help them to achieve them and find a pathway forward. The leader's role is thus at least in part defined by their role in the surrounding social structure of the organization (White and Shullman, 2010, p. 96). Consequently, learning agility is necessary for leaders to cope with change and to act effectively under conditions of uncertainty and ambiguity. White and Shullman (2010, p. 98) use the term 'bandwidth' as a concept for the ability to mitigate risk and indecisiveness, make greater use of inclusiveness, use delegation, consensus building, and brainstorming in coping with change.

Leadership Development

There is a general perception that leaders play an essential role in the operation of organizations and that leadership skills are complex and difficult to learn. As Riggio (2008, p. 383) points out, management training and leadership development is a very large business; many organizations spending a vast amount of resources providing for it.

While there is no universally accepted model of leadership development, most leadership and management training programs do have a positive impact on employees. Leader development is thus characterized by specific competencies performed at specific levels in given organizations and markets in order to leverage the organization towards efficiency and effectiveness. According to Riggio (2008, p. 384), leaders must possess a desire and readiness to learn and a commitment to lead.

There are some factors for leadership that seem to be universal. Leadership characteristics include emotional and social intelligence, reason and decision-making capacity, and the effective use of authority (Riggio, 2008, p. 385). A distinction can be made between leader development and leadership development. Leader development is a leader-building capacity to lead through skill acquisition, motivation, and awareness of others. Leadership development is concerned with the collective leadership capacity of the organization (Day, 2000). Riggio (2008, p. 386) suggests that organizations need to allow leaders to 'grow into' their roles by taking on organizational goals. He emphasizes the importance of 'trigger' or 'crucible' events – high impact experiences which define the leader's development.

Leadership Communication

Leadership communication styles, according to De Vries, Bakker-Pieper and Oostenveld (2009, p. 367) are related to four factors: knowledge-sharing behaviors, perceived leader performance, leadership satisfaction, and subordinate commitment. Leadership styles are divided mainly into either human-oriented (charismatic) or task-oriented leadership. The former are understood as more inherently communication-centered than the latter, but not exclusively so, and may be characterized by the difference between high and low communication contexts. De Vries et al. (2009) define leadership communication as a " ... distinctive set of interpersonal communication behaviors geared toward the optimization of hierarchical relationships in order to reach certain group or individual goals" (p. 368). Eight main factors of low to high communication contexts are defined by Hall (1976), and Gudykunst and Toomey (1988):

- Inferring meaning
- Indirect communication
- Interpersonal sensitivity

- Dramatic communication
- Use of feelings
- Openness
- Preciseness
- Positive perception of silence.

These factors take place on a continuum of interpersonal communication styles and intrapersonal thoughts and feelings with regard to communication. As De Vries et al. (2009) suggest, supportive communication styles usually result in greater satisfaction among employees, while dominant styles may produce less favorable outcomes both in clinical and educative settings. However, children's attentiveness may be stronger when communication is unambiguous and dominant (De Vries et al., 2009, p. 369). Between the two extremes of dominant and passive communication is the intermediate concept of knowledge-sharing, defined as the process whereby individuals "exchange their (tacit and explicit) knowledge and jointly create new knowledge" (De Vries et al., 2009, p. 369). Human-oriented leadership has inherently more relational content than task-oriented leadership and is characterized by a supportive communication style. This may be reflected in some aspects of leadership performance – an expressive style often has a greater effect than speech content (De Vries et al., 2009, p. 369).

Ethical Considerations in Leadership Positivity and Transgression

Ethical leadership may be associated with both transformational and transactional leadership, both in terms of the positive effect on organizational culture in the former, and the contingent reward dimension in the latter. If leadership is viewed as effective and ethical, employee satisfaction will be higher and employees may be willing to invest more effort in the enterprise. According to Ladkin (2008), effective and ethical leadership involves three dimensions: mastery of self and context, coherence between message and actions, and purpose acting assertively towards one's goal. According to Brown, Trevino, & Harrison (2005), ethical leadership is associated with: consideration behavior, honesty, trust in the leader, fairness in interactions, and socialized acceptability.

Ethical leadership is characterized by authentic self-awareness, relational transparency, a coherent moral perspective, and balanced processing (Toor & Ofori, 2009, p. 536). Interestingly, ethical leadership is sometimes negatively associated with laissez-faire leadership due to the tension between self-reliance and the directiveness of this leadership style. Servant leadership, in comparison, is rated more highly from an ethical viewpoint in that it places emphasis on the growth of subordinates. The ethical characteristics of three main leadership styles are given below (Toor & Ofori, 2009, p. 536).

Table 2. Characteristics of leadership styles

Servant leadership	Spiritual leadership	Transformational leadership
Self-awareness	Concern for others	Idealized influence
Authentic behavior	Integrity	Intellectual stimulation
Positive modeling	Role modeling	Inspirational motivation
Creating value for community	Altruism	Idealized consideration
Helping subordinates to succeed	Hope/faith	A culture of positivity
Conceptual skills	Emotional healing	Organizational commitment

Tumasjan, Strobel and Welpe (2011, p. 611) suggest there are four kinds of main influences on the ethical reasoning process: social, cultural, psychological and physical. They argue that social distance is an influence on how leadership ethics are rated. Social distance is the perceived gap between the actors of the moral issue and those evaluating it. This leads to 'unethical decisions' being evaluated more extremely in a higher psychological distance, whereas a lower social distance is likely to promote more consideration of the leader's motives.

It also accounts for the discrepancy between the ethical evaluation of leaders of one's own organization and that of leaders in general (Tumasjan et al., 2011, pp. 609-610). Overall, according to Toor et al. (2009, pp. 534-535), ethical leadership may be characterized by character, honesty, integrity, altruism, trustworthiness, collective motivation, encouragement, and justice. Arguably, ethical leadership should extend beyond the "demonstration of normatively appropriate conduct" (Brown et al., 2005) through interpersonal relationships and communication, reinforcement, and decision-making. This is to create an organizational climate which is conducive to positivity, growth, and the coherent interrelation and conduct of organizational objectives.

Leadership Behavior Theory

Larsson and Vinberg (2010, p. 317) suggest that the two traditional dimensions of leadership behavior theory are relation-orientation and structure-orientation, and that these operate in balance within an organization. However, a third perspective is that of contingency-management, in which leadership behavior changes with situational aspects and preferred outcomes (Larsson, & Vinberg, 2010, p. 318). Furthermore, these authors assert that while leadership can considerably affect organizational effectiveness, the relative strength of this effect varies. This is evidenced through the study of Mott (1972) who empirically tested for two leadership behavioral factors with the strongest effectiveness on outcomes (task-orientation and group maintenance).

Task-oriented behaviors are frequently used to improve organizational adaptation to external environments, whereas relation-oriented behaviors are used to improve internal environments, frequently human resources (Yukl, 2006). However, structure orientation in leadership style is often used to close the gap between a desired organizational performance state and the actual state. Correspondingly, it may be reduced when the necessary subordinate performance has been reached. Larsenn and Vinberg (2010) suggest that there are nine behavioral groupings along which leadership performance may be measured. These vary from the cognitively complex to maintaining everyday conversational presence (p. 329):

1. Strategic and visionary leadership role
2. Communication and information
3. Authority and responsibility

4. A learning culture
5. Maintaining worker conversations
6. Plainness and simplicity in communications
7. Emanating humanity and trust
8. A visible presence
9. Reflecting personal leadership.

Larsonn and Vinberg (2010) also suggest six primary factors of greatest leadership behavior influence on subordinate health and performance: 1) consideration, 2) the initiation of structure in stressful situations, 3) autonomy – allowing subordinates to control aspects of their work environments, 4) the production of meaning in the work task, 5) the provision of intellectual stimulation, and 6) a human-oriented and charismatic style (p. 320). While these factors may considerably influence organizational performance in with regard to leadership and subordinate relations, they take place in the context of a concern for organizational development and communication. A necessary feature also is the creation of an infrastructure for the flow of information and communication, for example, through flexible communication channels, regular meetings, and cross-functional discussions and dialogues (Larsson & Vinberg, 2010, p. 327). The effectiveness of leader-subordinate relations may also be influenced by the spectrum of authority and responsibility.

In a healthy organizational climate, each worker has ownership of specific tasks, areas of responsibility, and sufficient authority with which to conduct their work. The lack of penalties or punishments in many environments is seen as important, that is, there must be enough toleration of an error margin. Frequently incentivisation may enhance worker productivity. Plainness and simplicity in communication may also work to improve organizational efficiency by allowing decisions to be made under unambiguous conditions. Long-lasting employee health and a commitment to investment in employees are also seen as factors that may influence organizational performance (Larsson & Vinberg, 2010, p. 327). The visibility of leaders may (or may not) also influence employee productivity. Productivity may decrease in the absence of autocratic leaders but increase in the presence of consultative leaders.

The Five-Factor Model of Leadership

Seiler and Pfister (2009, p. 42) advocate the five-factor model of leadership behavior. It defines leadership behavior as the result of a combination of qualities, characteristics, performances and actions. These are described by the leader's, 1) overall competence, 2) ability to interact and lead a group, 3) knowledge and navigation of organizational rules, structures, and procedures, 4) knowledge of internal organizational contexts and external threats and opportunities, and, 5) ability to understand the requirements of any given imminent situation.

These five factors, along with their components and examples are presented in the table below (Seiler & Pfister, 2009, p. 44).

Table 3. The five-factor model of leadership

Factor	Components	Examples
Individual competence	Professional competence	Job knowledge, technological knowledge.
	Strategic competence	Strategic decision-making, knowledge management, problem-solving skills.
	Personal competence	Stress resistance, self-motivation, hierarchical awareness.
	Social competence	Empathy, tolerance, communication skills, leadership skills.
	Intercultural competence	Foreign language knowledge, knowledge of foreign cultures and countries.
Group	Structural aspects	Group composition, objectives and duties, norms and roles.
	Dynamic aspects	Relationships, communication, group dynamics.
Organizational	Strategy	Internationalization, expansion, downsizing, reward, salary, bonus systems, company goals.
	Structure	Hierarchical structure and responsibilities, infrastructure.

Factor	Components	Examples
	Processes	Standardized processes, knowledge transfer and management.
	Culture and climate	Feedback culture, error management culture, ethical climate.
Context	Static components	History, geography, national culture.
	Dynamic components	Political, economic, and social development; international treaties and law; weather.
Situation	Clarity	Information availability, ambiguity and relevance.
	Familiarity	Preliminary experiences with comparable situations, novelty of the situation.
	Pressure	Pressure for decision and action, time pressure, danger.

Destructive Leadership Behavior

According to Aasland, Stogstad, Notelaers, Nielsen and Einsuren (2010), constructive leadership is centered on behavior that supports and enhances the attainment of goals within an organization. This may involve motivating staff, promoting the well-being and job satisfaction of subordinates, and using organizational resources efficiently (Aasland et al., 2010, pp. 440-441). This may be achieved through both the transformational and transactive leadership paradigms. However, sometimes leadership may go wrong. The reasons for this may be manifold, but the influence of destructive leadership can be pervasive in an organization.

Aasland et al. (2010) define destructive leadership as a "systematic and repeated behavior by a leader, supervisor or manager that violates the legitimate interest of the organization by undermining and/or sabotaging the organization's goals, tasks, resources, and effectiveness and/or motivation, well-being or job satisfaction of subordinates" (p. 439). Thus whether leadership is constructive or destructive, it occurs on a continuum from 'anti' to 'pro' organization in its effects and outcomes. Leadership can be destructive in relation to the structural elements of the workplace, leading to difficulties in the performance of work tasks or with resources, or in terms of the human-orientation, creating difficulties in organizational communication or staff relations.

Destructive leadership is not defined by any one characteristic, but more frequently a variety or combination of behaviors which do not entirely reinforce organizational goals. It is an unpleasant aspect of many workplaces that at some point in their working lives, staff will be exposed to some form of negative leadership, be this "abusive supervision" or "petty tyranny" (Aasland et al., 2010, p. 439). Many of the traits of negative leadership may result from the dark side of positive leadership. Some example include: the charismatic leader who is unable to remain objective or who personalizes workplace issues; the flexible leader who is seen as Machiavellian in his or her dealings with staff; the leader who is pro-structure who becomes autocratic in his or her communications; or the leader who achieves great successes and who becomes narcissistic in his or her relations with others. Many of these negative traits may arise from the original issue of control and obedience in conditions of uncertainty.

The destructive leadership behavior model, outlined by Aasland et al. (2010, pp. 440-441) as a model for when leadership expresses negative traits, has four features. The first is constructive and three are more overtly negative. These last three are tyrannical, derailed, and supportive-disloyal leadership. The former is laissez-faire leadership, which is situated in a neutral middle between constructive and destructive leadership. Tyrannical leadership behavior is characterized by pro-organizational behavior coupled with anti-subordinate behavior. Results may be obtained by tyrannical leaders at the expense of subordinates and personal encounters may be characterized by humiliation, belittlement or manipulation over work tasks (Aasland et al., 2010, pp. 440-441). This may lead to a dissonance in the evaluation of the 'tyrannical' leader's behaviors – which may be seen as results-driven by superiors and as abusive by subordinates.

In comparison, derailed leadership may be characterized by anti-organizational and anti-subordinate behavior. In many ways this is the worst form of destructive leadership as it may involve bullying, manipulation, or deceptive behavior, and also the theft of organizational resources. Supportive-disloyal leadership combines both positive and negative features. This style of leadership is characterized by the motivation and support of subordinates while at the same time either using organizational resources inappropriately or in unethical ways. At its most destructive it results in the 'pied-piper' phenomenon with the leader encouraging subordinates in activities that are against their own or the organization's interests.

Figure 1. A model of destructive leadership behaviour (Aasland et al., 2010, p. 440)

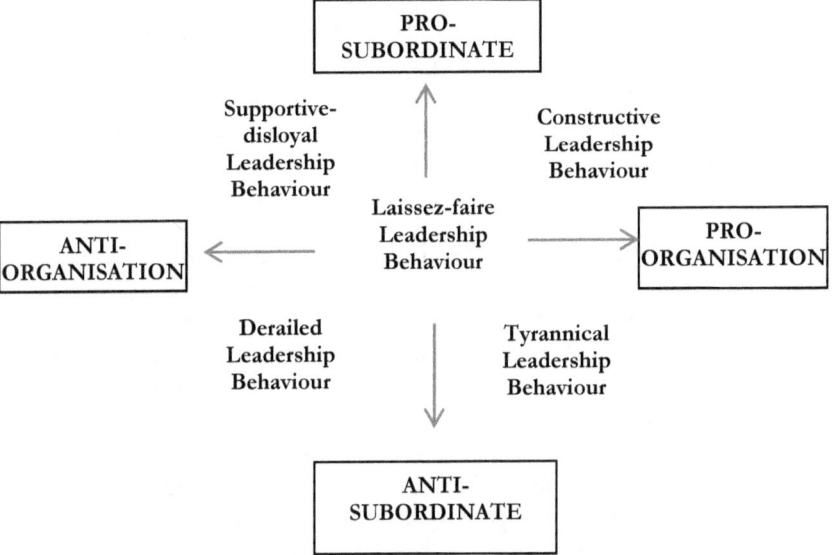

Wisdom and Leadership

In contemporary organizations, the concept of inspired leadership is more often linked, or even subordinated, to efficiency in the attainment of organizational performance/profits than to the concept of wisdom. Wisdom differs from such utilitarian rationale in so much as it is concerned with the attainment of good lives. Consequently, in organizations run primarily on business principles, wisdom is not often promoted as an organizational goal, or, if it is, only when the business operational environment has achieved its main objectives or output targets. However, according to Yang (2011, p. 617), wise leaders have a clear vision of good life, embody this vision through their behavior, demonstrate compassion, and also have an infectious ability to create this good life for others.

Wise leadership is both a product of personal integrity and action in the organizational environment resulting from either higher-level reasoning, a collective system of practical knowledge, or of knowledge acquired through leadership experiences (Yang, 2011). Thus wisdom involves learned competencies, but also the application of qualities such as intelligence and creativity in the context of organizational objectives. Sternberg (2007, p. 38) defines wisdom as,

> The use of successful intelligence, creativity, and knowledge as mediated values to a.) seek to reach a common good, b.) balance intrapersonal (one's own), interpersonal (others), and extra personal (organizational, institutional, and/or spiritual) interests, c.) over either [the] short [term] or long term to, d.) adapt to, shape, and select environments. (p. 617)

For McNamee (1998), wisdom is seen more as a communal knowledge-creation activity that is engendered through organizational communication and which emerges through the process of organizational activity. It is thus a form of distributed cognition. Similarly, for Sampson (1998) wisdom is understood as the properties of certain kinds of enduring conversation (cited in Yang, 2011, p. 619).

Thus wisdom may be an organizational variable that arises in situations when good leaders exercise their judgment about people and resources efficiently and humanely. Wisdom thus may be a relativistic quality that emerges as the result of learned knowledge in dealing with organizational or psychosocial crises (Yang, 2011, p. 618). It involves both cognitive and pragmatic qualities, and can be seen as "accumulated knowledge of fundamental pragmatics of life" (Yang, 2011, p. 619). Traits of wise leadership include (Yang, 2011, p. 617):

- Initiative in social situations
- Responsibility in task completion
- Toleration of frustration and delay
- Acceptance of consequences of action
- Ability to absorb interpersonal stress
- Ability to influence behavior
- Ability to structure relevant social systems

- A sense of self-confidence and personal identity
- Originality in problem-solving
- Vigor and persistence in pursuing goals
- Humility and strong professional will
- A strong sense of ethics and fairness.

Thus organizational knowledge and an integrated personality is not itself sufficient for the exercise of leadership wisdom; rather it also involves practical application of purposeful behaviors. Traditionally, wisdom has not always been associated with the business environment, but has rather been the preserve of historical writings. Leaders may or may not be the conduits for such knowledge. At this level, wisdom is more about leading a 'good life' than about the execution of business or organizational tasks. This does not mean leading an easy life without adversity, but rather by building personal attributes through character development.

Yang (2010, p. 619) discerns five components to these attributes: 1) factual knowledge about lifespan development, 2) procedural knowledge about how to deal with life's problems (involving critical reflection of past events), 3) rich knowledge about contexts of life and dynamics, 4) knowledge about the relativism of life's goals, and 5) management of uncertainty. Thus a distinction can be made between overall wisdom knowledge, which is concerned with life in general, and leadership-knowledge which is more about the accomplishment of organizational goals. For a leader to exhibit wisdom within an organization, he or she would engage in careful observation, be prepared to take the subjective into account, value humane outcomes, take practical action in everyday life, and be articulate and understand the virtues of social reward (Yang, 2011, p. 620).

Servant Leadership

As Searle and Barbuto (2011) suggest, servant leadership is a strongly positive leadership style which differs from other leadership styles in so much as it focuses on the needs of others in the organization. The emphasis in this style of leadership on service over self-interest implies that it is rooted firmly in an altruistic philosophy (Searle & Barbuto, 2011, p. 107).

Figure 2. Positive variables in servant leadership (After Searle & Barbuto, 2011, p. 109)

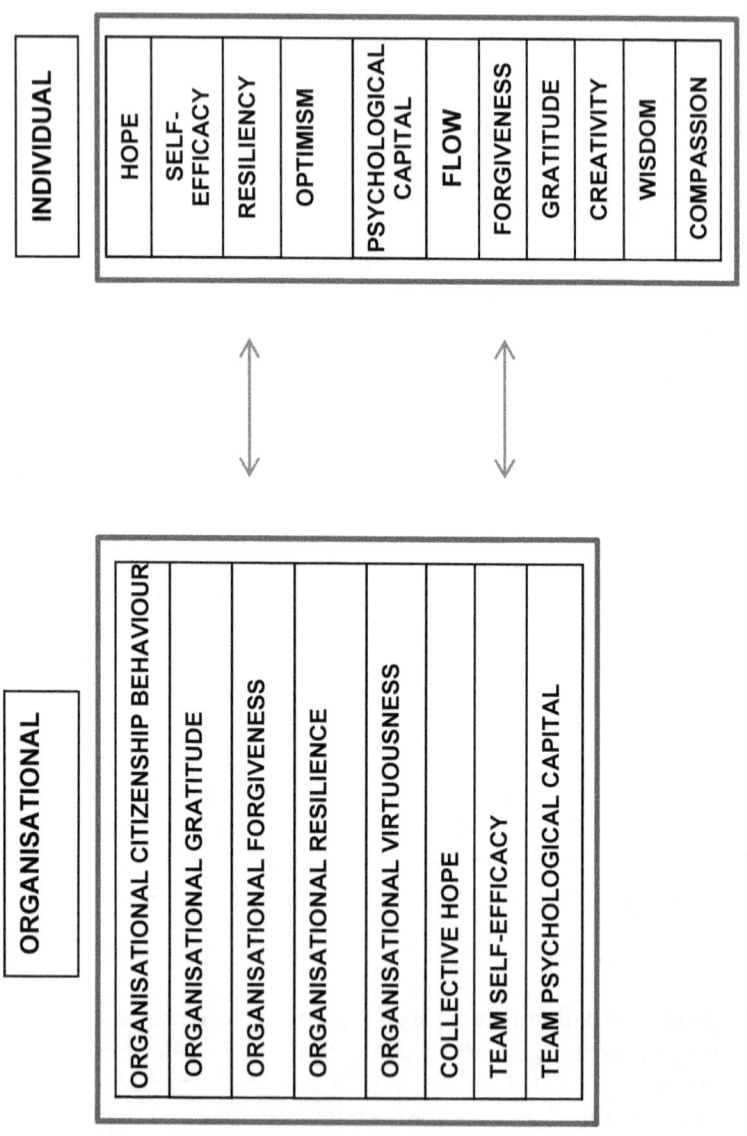

According to Spears (1995) there are five main dimensions along which servant leadership is expressed: altruistic calling, emotional healing, wisdom, persuasive mapping, and organizational stewardship. The first of these five factors is concerned with a fundamental choice to serve others. Emotional healing is the ability on the part of the servant leader to recognize the need for and facilitate a healing process. This ability to inspire recovery from trauma requires empathy and sensitivity towards others.

The wisdom of servant leaders is describes as an ability to exercise sound judgment and provide practical solutions to problems, while being mindful of implications and outcomes which are directed towards maximizing the ability of others to lead good lives. Persuasive mapping is an ability to use mental models and rational decision-making and to encourage lateral thinking in others. Organizational stewardship is described as the preparation of the organization to make a positive contribution to society. In many organizations, this can be accomplished through either community development programs or outreach activities.

Searle and Barbuto (2011, p. 111) argue that servant leadership is most effective when an organization has previously been under a regime of less ethical orientation; indeed organizational virtuousness is positively related to performance. Furthermore, organizational resiliency and compassion are qualities which underlie the application of the service provided by servant leaders. Servant leadership may differ notably from other forms of leadership in that it does not place foremost priority on tangible outcomes (profits and material gains), although these may be a by-product of it. Rather, the focus is on the instillation of a culture of positivity throughout the organization, more akin to the stewardship approach to resources and people management. Foremost in the servant leader's repertoire of positive actions is the desire to support a caring and co-creative environment.

Servant leaders may also be cognizant of 'hope theory,' which consists of an analysis of goal routes and motivations in assailing the route to the desired outcome (Searle & Barbuto, 2011, p. 112). Arguably, those who pay critical attention to these two factors will be more effective in attaining their goals through an ability to foresee and overcome obstacles. This can be aided by the servant leader's pervasive mapping skills – the articulation of a mental model which provides pathways for followers to achieve goals, aims and objectives (Searle & Barbuto, 2011, p. 113).

A servant leader must also be aware of the motivational factors of organizational performance or willpower. Organizational blocks may be overcome by the servant leader's ability to listen, empathize, and suggest alternative routes. Hope is itself a positive emotion which encourages adaptation, success, and optimism and it can overcome apathy and loss of motivation. Extending from the servant leaders' program is the instillation of a form of organizational virtuousness, which is transcendental in its effects (Searle & Barbuto, 2011, p. 114). This may be achieved by the implementation of policies which create value for the organization by contributing to community involvement and staff and stakeholder betterment.

Self-esteem: Definitions

Whilst there is considerable debate about what self-esteem is, how it should be defined, and how it is measured, the main components are ostensibly the extent to which a person understands their own ability and the extent to which a person is at ease with themselves and others. As Koch and Shepperd (2008) suggest, contemporary definitions of self-esteem regard it as a "global, unitary evaluation of the self" (p. 55). It can be understood as something more than self-competence and self-liking, involving non self-referential dimensions of feelings of being accepted by and included with others. Thus it is a "monitor of inclusionary status" (Koch & Shepperd, 2008, p. 55). The definition according to Newstrom, Gardner and Pierce (1999) is that "self–esteem is a self-evaluation reflecting the extent to which individuals believe themselves to be capable, significant, successful and worthy" (p. 19). High self-esteem is more than just a feeling of being accepted and included by others, and low self-esteem is the consequence of being rejected by others as any dimension of human behavior and perception may be affected by other dimensions at the same time and with different degrees of intensity.

Although the attempt to shield people from negative performance may be detrimental to long-term growth and development and to formulations of normalizing tendencies, it can occur in the absence of high competence. In fact, many comedians base much of their humor routines on the awkward juxtapositions of perceived low competence. However, this is likely to work less well in other areas of life. Certainly in educational settings, concerns with esteem preservation may be pervasive as it is inherently related to the motivations for educational pedagogy. However, as with different sections of society, different cultures may have different views of self-worth. These may be expressed in complex relationships of inclusion/exclusion, participation and retreat. Common to all is a concern with identity construction and maintenance of self and other perception.

As such, self-esteem involves a personal assessment of 'worthiness.' However this may be expressed as a feeling of approval, elation, or as more of a long-term mood-related concept that reflects the extent to which the individual likes or dislikes him or herself. Of course this is complicated by the fact that people are multi-faceted. Each individual may have parts of themselves which they like (or don't like) and parts they like more than others. Some of the main determinants of self-esteem are the extent to which messages received from others are enhancing or deprecating. Individuals are also affected by messages received at work and the extent to which these are affirming, validating, accepting, or disconfirming. People's sense of self-esteem is also affected by the 'life-system' or ecosystem to which they are exposed, the physical environment by which they are shaped, and also the level of competency they experience as derived from personal experience (Newstrom, Gardner & Pierce, 1999, p. 19).

As Newstrom et al. (1999) suggest, self-esteem is a self-perception. It thus has intentional qualities involving beliefs about other's beliefs about self. Self-esteem is important, not only for self-identity, but also as a component of the workplace. It may shape a person's orientation to work, their capacity for self-direction and self-control, and their motivation and ability to engage in meaningful work-related activities.

Organizational Self-Esteem

Cooperative behavior in organizations is described as 'organizational citizenship.' Although it is frequently not rewarded it has positive consequences for the organization. The natural extension of organizational cooperation is called collectivism. Van Dyne, Vanderwalle, Kostova, Latham, and Cummings (2000) explain that collectivism involves an individual's belief that "collective or group interests should take precedence over individual self-interest" (p. 5). There is some evidence that people with a collectivist orientation derive more of a sense of self from the organization and may contribute to the collective by subordinating personal interests to organizational interests (Van Dyne et al., 2000, p. 5). Collectivism may also inspire interpersonal helping (Moorman & Blakely, 1995).

Organizational cooperation should also be related to trust. Van Dyne et al. (2000) describe trust as a "stable individual difference" defined by a "generalized expectancy to attribute benevolent intent to others and rely on information received from others about uncertain environmental states and their outcomes in situations involving risk" (p. 6). Trust becomes increasingly relevant under conditions of uncertainty. Trust is related to fairness and the idea that positive acts will be reciprocated. All three variables may inform the level of organizational cooperation. Hence, the derivation of the individual's positive self-concept is derived from favorable experiences. Trust is also relevant in non-work settings where inter-personal exchanges are frequent.

Propensity to trust has a positive relationship with organizational citizenship. However self-esteem within an organization is complex as there are more constituent features to interact with the complexity of personal self-esteem. As Van Dyne et al. (2000) suggest, "Global self-esteem reflects a person's overall evaluation of the self; it is multidimensional and composed of task and situation-specific components" (p. 7). Given that self-esteem is inherently concerned with a person's sense of self, and while the culture and norms of the organization may all be influenced by cooperation, trust and collectivism, it is doubtful if the work environment should be responsible for a person's global self-esteem.

However, when a person experiences these factors in congruence within an organization they are more likely to experience higher self-esteem within the organization as a consequence. The person's sense of self is influenced by individual characteristics and their expression in particular circumstances. However, people who favor collectivist dispositions may have an independent idea of self that mediates between a sense of individuality and a sense of group membership. Thus, self-esteem within organizations may be a relational concept. Trust may act as a form of social bond that increases the likelihood of favorable experiences within the organization. On the other hand, organizational members who do not develop common bonds of trust are less likely to derive positive feelings of self-worth from the organization. Reciprocity or exchange may lead to self-validation from others; furthermore trust leads to sustained cooperation on the basis of reciprocity.

There is also some evidence that tenure (length of association with the organization) not only leads to a better understanding of organizational practices, but also to the role of the self within the organization. Van Dyne et al. (2000) affirm the importance of ongoing relationships and principles of social exchange; thus, organizational tenure also influences the quality of intrinsic motivation. On the other hand, for those who have negative experiences within the organization on an ongoing basis, these are more likely to increase over time and lead to a decrease in organizational involvement (Van Dyne et al., 2000, p. 10). Consequently collectivism and trust will be more strongly correlated to organizational self-esteem for individuals with longer associations with the organization (Van Dyne et al., 2000, p. 10).

As Bowling, Eschelman, Wang, Kirkendall and Alarcan (2010) suggest, self-esteem can also play a role in predicting attitudes and behaviors. A related question is: should self-esteem be conceptualized as a multivalent construct where individuals may have multiple perceptions of self-worth across a variety of different life domains? Different spheres of life may have differing perceptions of self-worth valence for different people. As Bowling et al. (2010) suggest, general self-concepts of self-esteem differ from organizational-based self-esteem because they refer to an individual's belief of her or his self-worth as an 'organizational member.' These authors also posit that positive relationships will exist between organization-based self-esteem, general self-esteem, self-efficacy, and emotional stability (Bowling et al., 2010, p. 601). Furthermore, at the level of organizational membership there will be positive correlations between organizational-based self-esteem and these six factors (p. 601):

1. Job complexity
2. Autonomy
3. Effective leader behaviors
4. Social and organizational support
5. Psychological ownership
6. Salary

However, negative correlations exist between organizational-based self-esteem and job stressors such as (p. 601):

- Role ambiguity
- Role conflict
- Role overload
- Job insecurity

Also, organization-based self–esteem is positively correlated with job satisfaction, effective and normative organizational commitment, and job involvement, however it is negatively correlated with turnover intentions (Bowling, 2010, p. 601). Although organization-based self-esteem provides stronger relationships with work-related variables, general self-esteem has a stronger relationship to employee health (Bowling, 2010, p. 601).

Organizational self-esteem takes place in a context of both work and family. Attitudes to work-to-family conflict and well-being/job attitude relationships are influenced by the degree of family supportive climate as well as the degree of job control. These in turn may affect organization-based self-esteem (OBSE). As Mauno, Kinnunen and Ruokolainen (2006, p. 210) suggest, job demands (time and effort/work-to-family conflict) are more commonly associated with strain-based outcomes (physical symptoms), whereas job resources (job control, family support, organization-based self-esteem) are correlated to motivational factors (job satisfaction, commitment to an organization).

However, Mauno et al. (2006) also suggest that work-to-family conflict is a prevalent factor in most western workplaces. The reasons they suggest for this include market-driven globalization and the continuous demand for growth and efficiency. Responding to economic realities in dynamic organizations means varying workloads that may encroach on family life (Mauno et al., 2006, p. 210). There is a reciprocal action or 'vicious cycle' between work-to-family conflict and stress which may lead to negative outcomes, impaired performance, diminished well-being, and eroding attitudes.

Nevertheless there are some resources for the stressed individual which may buffer them from this negative outcome. They generally fall into two ranges: first, various individual factors (personality, attitude or dispositional) and organization-based outcomes (e.g. job control, family supportive climate). Both of the latter may play a moderating role on the effects of work-to-family conflict on well-being (Mauno et al., 2006, p. 210). Job demands may result in strain reactions, such as stress, burnout, and ill-health, whereas lack of resources (e.g. diminished job control or social support) may lead to less effective goal accomplishment.

The resultant emotion may be frustration, fear of failure, or at the extreme, despair. The latter may cause negative coping behaviors such as withdrawal and reduced organizational commitment to job satisfaction (Mauno et al., 2006, p. 212). Thus, work-to-family conflict may be positively correlated to physical symptoms and negatively related to job satisfaction (Mauno et al., 2006, p. 213). Work-to-family conflict may also show a positive correlation with strain (physical symptoms) than with attitudinal outcomes (job satisfaction, organizational commitment). Similarly, poor job control may result in impaired health and job attitudes, whereas high job control is more likely to operate as a resource which produces positive outcomes. High job control may also alleviate negative job stressors as well as improve well-being outcomes.

There is also some evidence that work task autonomy may buffer against family-to-work interference by lessoning the effect of two negative reactions – exhaustion and cynicism (Mauno et al., 2006, p. 213). On the other hand, a supportive family environment has a positive effect on organizational commitment and job satisfaction (Mauno et al., 2006, p. 214). Mauno et al. (2006) describe work-to-family conflict as "bi-directional as it may lead either way – the one to the other and originate in either sphere" (p. 212). However work-to-family conflict is more common than family-to-work conflict. Greenhaus and Bautell (1985) distinguish between time-based conflict, strain-based conflict, and behavior-based conflict. In the former, the time demands of one role interfere with the performance of another role. In strain-based conflict, strain or fatigue hinders the performance or resources available for another role. In behavior-based conflict, behavioral styles in one role are incompatible with the behaviors in another role – leading to values conflict (Mauno et al., 2006, p. 212).

Newstrom et al. (1999) argue that self-perceptions of self-efficacy can greatly affect motivation, attitudes and work-related behaviors. Self-esteem is also a powerful determinant of work-place orientation, capacity for self-direction, and self-control (Newstrom et al., 1999, p. 9). As they suggest, employees with high task-based self-esteem on the whole perform better and with greater satisfaction than those with low task-based self-esteem. High task-based self-esteem may lead to a positive cycle of reward, confidence, and competence, whereas those with lower task-based self-esteem may limit their success through predictions of variable success (Newstrom et al., 1999, p. 10).

An individual with strong self-esteem may internalize the belief that they are an important part of the workplace. Positive feedback may also influence workplace success. If supervisors communicate to employees that they are valued and trustworthy, a reciprocation of positive work attitudes, motivations, and performance are likely to follow leading to greater overall self-esteem (Newstrom et al., 1999, p. 11). A culture of respect may then be created. As Newstrom et al. (1999, p. 12) state, other factors that may impact positively on organizational climate and self-esteem include: enriched work, varied work tasks, accurate timely feedback, worker autonomy, reaffirmation of value, trust for task completion, and promotion based on merit. However, these need to be sustained by adequate resources and co-worker support.

Self-Efficacy, Self-Esteem and Self and Identity

As Van Der Roest and Kleiner (2011, p. 26) suggest, self-efficacy theory may reflect the role that human biology plays in confidence and motivation. In organizational psychology, it is not necessarily the case that an employee's confidence level is dictated by a manager's ability to inspire. Hindrances to self-efficacy include workplace inequalities or responsibilities that are in excess of abilities. However, self-efficacy is a quality that can be improved and to some extent learned, and need not necessarily be seen as a side-effect of self-confidence (Van Der Roest & Kleiner, 2011, p. 27). If self-efficacy is a quality that the individual person may improve on, it is also one that management can improve on. Self-efficacy is amenable to improvement in both physical and mental senses. Albert Bandura (1994) proposed four criteria for improving self-efficacy. These techniques include inactive mastery, vicarious modeling, verbal persuasion, and arousal.

Van Der Roest and Kleiner (2011) suggest that protein-rich and carbohydrate-low diets may also improve self-confidence and performance. This is because protein-rich food is made up of derivatives of neurotransmitters and amino acids, which build serotonin (a biochemical agent that improves confidence). Diet and regular exercise it is seen as beneficial. Sustained cardio-vascular activity may release endorphins which disinhibit dopamine neural pathways. As they suggest, "the balance between dopamine and glutamine is the primary determinant in the bio-chemical reactions that control self-efficacy" (Van Der Roest & Kleiner, 2011, p. 30). Coupled with cardio-vascular activity, an increase in alpha-wave activity (bHz and above) increases levels of serotonin). Serotonin also improves long-term memory and may lower stress and anxiety. While dietary nutrition is usually associated with appearance and physical health, it also greatly affects mental health: amino acids build the neurotransmitters that define personality, behavior, and cognition (Van Der Roest & Kleiner, 2011, p. 32). Thus, as Van Der Roest and Kleiner suggest, the factors which influence self-efficacy are multi-dimensional and include biological as well as social, cultural, and workplace constituents (p. 35).

Personality and Self-Esteem

As Neustadt, Chamorro, Premuzic and Furnham (2006) explain, higher self-esteem results from secure work attachment and lower self-esteem from insecure attachment. Interestingly, extroversion and agreeableness are also positively correlated with secure attachment in the working environment (Neustadt et al., 2006, p. 209). The distinction between a person's conception of their 'ideal self' and their 'actual self' also has a bearing on self-esteem. Those whose 'actual-ideal self' conceptions were more divided generally have lower self-esteem (Renaud & McConnell, 2007, p. 41). Here two related theories of personality are relevant. The first is entity theory, which argues that personality is relatively fixed and unchangeable. The second is the incremental theory of personality, which posits that personality is malleable and flexible (Renaud & McConnell, 2007, p. 41). Those whose 'actual self' and 'ideal self' conceptions are divided, and those who also believe that the entity theory of personality is more relevant, will also report lower self-esteem.

However, the incremental theory of personality leaves open the possibility for change adaptation and improvement – the closing of the gap between actual and ideal selves. Self-awareness and self-regulation may also be affected by the discrepancy between actual and ideal selves. Goal-directed behaviors may seek to close the gap between perceived and desired states of being. Arguably beliefs about the possibility to change may mediate the effects of large discrepancies (Renaud & McConnell, 2007, p. 42).

The major factors influencing personality at work include the extent to which one is secure/autonomous and insecure/dependent. Insecurity is significantly correlated to neuroticism (negatively) and openness to experience positively correlated with attachment. Self-esteem is positively correlated with secure attachment at work (Neustadt et al., 2006, p. 209). Attachment relationships which are formed in early childhood may extend throughout life, as according to Neustadt et al. (2006), "attachment security promotes the development of a stable, positive self-image" (p. 209).

References

Aasland, M. S., Skogstad, A., Notelaers, G., Nielsen, M. B. & Einarsen, S. (2010). 'The prevalence of destructive leadership behavior'. *British Journal of Management*, *21*, 438-452.

Bandura, A. (1994). 'Self-efficacy'. In V. S. Ramachaudran (Ed.), *Encyclopedia of human behavior* (Vol. 4, pp. 71-81). New York: Academic Press.

Bowling, N. A., Eschelman, K. J., Wang, Q., Kirkendall, C. & Alarcon, G. A. (2010). 'Meta-analysis of the predictors and consequences of organization-based Self-esteem'. *Journal of Occupational and Organizational Psychology*, *83*, 601-626.

Brown, M. E., Trevino, L. K., Harrison, D. (2005). 'Ethical Leadership: A Social Learning Perspective for Construct Development and Testing', *Organizational Behavior and Human Decision Processes*, *97*, 117-134.

Corrigan, O. (1999). *Shakespeare on management*. London: Kogan Page.

Day, D. V. (2000). 'Leadership development: A review in context'. *The Leadership Quarterly*, *11*, 581-613.

De Vries, R. E., Bakker-Pieper, A. & Oostenveld, W. (2009). 'Leadership = communication? The relations of leaders' communication styles with leadership styles, knowledge sharing and leadership outcomes'. *Journal of Business Psychology*, *25*, 367-380.

Gavetti, G. (2011). 'The new psychology of strategic leadership'. *Harvard Business Review*. Prod. #: R1107K-PDF-ENG

Gudykunst, W. B., & Toomey, T. S. (1988). 'Culture and affective communication'. *American Behavioural Scientist*, *31* (3), 384-400.

Hall, E. T. (1976). *Beyond culture*. New York: Doubleday.

Hodgson, P., & White, R. P. (2001). *Relax, its only uncertainty*. London: Financial Times Prentice Hall.

Koch, E. J. & Shepperd, J. A. (2008). 'Testing competence and acceptance explanations of self-esteem'. *Self and Identity*, *7* (1), 54-74.

Ladkin, D. (2008). 'Leading beautifully: How mastery, congruence and purpose create the aesthetic of embodied leadership practice'. *The Leadership Quarterly*, 19, 31-41.

Larsson, J., & Vinberg, S. (2010). "Leadership behavior in successful organizations: Universal or situation-dependent?' *Total Quality Management, 21* (3), 317-334.

McNamee, S. (1988). 'Reinscribing organizational wisdom and courage: The relational engaged organization'. In S. Srivastva & D. L. Cooperrider (Eds), *Organizational wisdom and executive courage* (pp. 101-117). San Francisco: The New Lexington Press.

Mauno, S., Kinnunen, & Ruokolainen, M. (2006). 'Exploring work-and organization-based resources as moderators between work-family conflict, well-being, and job attitudes'. *Work & Stress, 20* (3),. 210-233.

Moorman, R. H., & Blakely, G. L. (1995). 'Individualism-collectivism as an individual difference predictor of organizational citizenship behavior'. *Journal of Organizational Behavior, 16,* 127-142.

Mott, P. E. (1972). *The characteristics of effective organizations.* New York: Harper & Row.

Neustadt, E., Chamorro-Premuzic, T. & Furnham, A. (2006). 'The relationship between personality traits, self-esteem, and attachment at work'. *Journal of Individual Differences, 27* (4), 208-217.

Newstrom, J., Gardner, D. & Pierce, J.(2007). 'A neglected supervisory role: building self-esteem at work'. *Supervision, 68,* 3, 9-12.

Rayner, S. (2009). 'Educational diversity and learning leadership: A proposition, some principles and a model of inclusive leadership'. *Educational Review, 61* (4), 433-447.

Renaud, J. M. & McConnell, A. R. (2007). 'Wanting to be better but thinking you can't: Implicit theories of personality moderate the impact of self-discrepancies on self-esteem'. *Self and Identity, 6,* 41-50.

Riggio, R. E. (2008). 'Leadership development: The current state and future expectations', *Consulting Psychology Journal: Practice and Research,* 60 (4), 383-392.

Sampson, E. E. (1998). 'The political organization of wisdom and courage'. In S. Srivasta & D. L. Cooperidder (Eds.), *Organizational wisdom and executive courage* (pp. 118-133). San Francisco: The New Lexington Press.

Searle, T. P. & Barbuto, J. E. (2011). 'Servant leadership, hope, and organizational virtuousness: A framework exploring positive micro and macro behaviors and performance impact'. *Journal of Leadership & Organizational Studies, 18* (1), 107-117.

Seiler, S. & Pfister, A. C. (2009). 'Why did I do this?: Understanding leadership behavior through a dynamic five-factor model of leadership'. *Journal of Leadership Studies*, *3* (3), 41-52.

Sternberg, R. G. (2007). 'A system model of leadership'. *American Psychologist*, *62* (1), 34-42.

Toor, S-u-R. & Ofori, G. (2009). 'Ethical leadership: Examining the relationships with full range leadership models, employee outcomes, and organizational culture'. *Journal of Business Ethics*, *90*, 533-547.

Tumasjan, A., Strobel, M. & Welpe, I. (2011). 'Ethical leadership evaluations after moral transgression: Social distance makes the difference'. *Journal of Business Ethics*, *99*, 609-622.

Van Der Roest, D. & Kleiner, K. (2011). 'Self-efficacy: The biology of confidence'. *Culture & Religion Review Journal*, *1*, 26-36.

Van Dyne, L., Vanderwalle, D., Kostova, T., Latham, M. E. & Cummings, L. L. (2000). 'Collectivism, propensity to trust and self-esteem as predictors of organizational citizenship in a non-work setting'. *Journal of Organizational Behavior*, *21*, 3-23.

White, R. P. & Shullman, S. D. L. (2000). 'Acceptance of uncertainty as an indicator of effective leadership'. *Consulting Psychology Journal*, *62* (2), 94-104.

Yang, S-Y. (2011). 'Wisdom displayed through leadership: Exploring leadership-related wisdom'. *The Leadership Quarterly*, *22*, 616-632.

Yukl, G. (2006). *Leadership in organizations* (6th ed). London: Prentice Hall.

ORGANIZATIONAL COMMUNICATION
Decision-Making and Workplace Trust

Decisions are a part of daily life, and many of them that we make are ordinary and commonplace. However, some workplace decision-making requires more complex thought. As D'Ambrosio (n.d.) suggests, there are three elements to any decision problem. First it may involve beliefs about the world, secondly, a set of action alternatives, and thirdly, the identification of preferences over possible outcomes of alternative actions (para. 3).

The law takes an interest in decisions which involve exchanges of property or which produce actions with harmful consequences (Mackenzie & Watts, 2011, p. 43). Mental capacity is a concept which describes the ability of a person to make rational autonomous decisions. Capacity refers to an ability or power to do something specific – to perform a task or to make a decision. In many Western cultures, mental capacity is tied to the concepts of ability, competence, and understanding (insofar as it implies a capacity to understand and appreciate the nature and consequences of a decision). 'Capacity' is a clinical term and 'competence' a legal term, but they are frequently used inter-changeably. However, decision-making is further distinguished by rational decision-making, which is making a decision based on sound cognitive principles. In the medico-legal context, mental capacity is understood in relation to a decision or act (Owen, Freyenhagen, Richardson & Hotopf, 2009, p. 81). Elliot (1991) regards capable decisions as being those which we legitimately hold to account.

Decisions about mental capacity are usually reserved for clinical settings and only qualitatively and peripherally in educational settings. There are two main kinds of capacity assessment: organic psychiatric disorders (learning disability and other organic brain syndromes) and non-organic psychiatric disorders (psychotic illness, depression, anorexia nervosa). The first category involves cognitive incapacity and the second incapacity involves non-cognitive terms. As Owen et al. (2009, p. 81) point out, in law, surrogate decision-making has been legitimized on the basis of incapacity as understood in cognitive terms, whereas surrogate decision-making in psychiatric settings is on the basis of risk or harm to patients or others (made because of a behavioral-mental evaluation).

The concept of mental capacity is linked to that of autonomy within legal theory, that is, a minimal level is required to be autonomous. Self-government is closely linked to autonomy, as is the concept of rational decision-making. There is also some debate over whether the concept of autonomy is value-laden (Owen et al., 2009, p. 82). For example, is it possible to distinguish between moral autonomy and psychological autonomy? Moral autonomy involves being able to give oneself freely (or to subject oneself) to norms or values. Personal autonomy is understood as the capacity to govern oneself according to one's own reasons or desires, but may also include aesthetic or etiquette-based values (Owen et al., 2009, p. 82).

There are also varying opinions on the nature and extent to which autonomy is rational. In the Kantian conception, autonomy is moral and rational, but another view holds that it is neither inherently moral nor rational. Yet another account holds that moral autonomy may (or may not) be rational and depends on behavior. How might this notion of autonomy be tested? The medical setting provides an example in which some treatments require patients' consent that may entail right of refusal. Ignoring this right would be to ignore a patient's autonomy or self-determination. However, the patient must also possess the understood 'mental capacity' to refuse, because only under such a condition can the refusal be regarded as autonomous. But this exercise of capacity acquires no moral significance. It is seen as a necessary condition for self-determination and the law operates in presumption of favor of capacity. Consent involves the ability to express a choice about treatment, understand information relevant to the treatment decision, appreciate the significance of the treatment, reason with relevant information so as to engage in a logical process of assessing treatment options, and communicate a decision.

Capacity tests are made to measure mental competence such as the MacArthur Competence Assessment Tool-Treatment (MacCAT-T). The concept of understanding is assessed by asking subjects to describe the nature of the information disclosed and questioning to check the level of understanding. Appreciation is assessed by gauging how personalized and context-specific the information is. Reasoning is evaluated by examining the subject's decision-making process, and weightings are ascribed to relevant information components (MacKenzie & Watts, 2011, p. 47). Clinicians thus surmise that decision-making capacity is dimensional, rather than categorical, as some dimensions may be present while others are absent.

Does justifying the risk-relativity of capacity assessments on the grounds of protecting patients from harm compromise autonomy? Owen et al. (2009) suggest that risky decisions are cognitively more demanding and seek a more stringent framework for assessing outcomes. But are decisions that involve risks to others necessarily cognitively harder? They certainly may be profound, but the level of information manipulation may not necessarily be more complex than less risky decisions. Formulating risk relativity may also involve a margin of error for capacity determination. As Owen et al. (2009, p. 99) point out, to judge incapacity falsely may result in well-intentioned treatment, but may result in harm that was preventable that was not autonomously chosen by the patient.

Coping With Stress and Conflict in Decision-Making

According to Bouckenooghe, Vanderheyden, Mestdagh and Van Laethem (2007, p. 605), there are four patterns of behavior associated with coping with stress and conflict in decision-making: a) to engage in vigilant information searching and trying to solve the problem immediately; b) become hyper-vigilant and panicking in the search for a solution; c) transfer the responsibility; and d) escape conflict through procrastination. Whatever course of action is chosen will depend on the cognitive level of the task and the amount of energy and resources the decision-maker is willing to invest in the problem (Bouckenooghe et al., 2007, p. 606).

Janis and Mann's (1977) conflict theory of decision-making holds that intense conflicts are likely to arise when a person makes an important decision. Conflict arises when there are simultaneous opposing tendencies within an individual when deciding to take a course of action. Robert Frost captured this conflict famously in his poem "The Road Not Taken" (1920):

TWO roads diverged in a yellow wood,
And sorry I could not travel both
And being one traveller, long I stood
And looked down one as far as I could
To where it bent in the undergrowth;

Then took the other, as just as fair,
And having perhaps the better claim,
Because it was grassy and wanted wear;
Though as for that the passing there
Had worn them really about the same,

And both that morning equally lay
In leaves no step had trodden black.
Oh, I kept the first for another day!
Yet knowing how way leads on to way,
I doubted if I should ever come back.

I shall be telling this with a sigh
Somewhere ages and ages hence:
Two roads diverged in a wood, and I —
I took the one less travelled by,
And that has made all the difference.

Feelings that the decision-maker may hold in situations of choice conflict include, emotional stress, hesitation, vacillation, and uncertainty (Bouckenooghe et al, 2007, p. 606). There are also considered to be four patterns of defective decision-making that can be maladaptive depending on its severity. These are unconflicted adherence (maintaining course of action), unconflicted change (uncritically adopting a new course of action), defensive avoidance (escaping), and hyper-vigilance (searching in a panic for a way out of dilemmas) (Bouckenooghe et al., 2007, pp. 606-607).

However, in international management of conflict, the negotiator (who must mediate between the choices accepted or declined by two or more parties) is frequently a transient agent, whose aim is to affect a comprehensive outcome to resolve the disputes of conflicted parties and do so in a way that restores values which are lost in conflict. Strategic thinking involved in such negotiattion also needs to include contingencies. The following parable illustrates the complexity of such negotiations:

> You are in a country where there are only liars and truth tellers, and they cannot be told apart by sight. You set out on a dangerous trip, because there is a fork in the road ahead that can lead to either danger or safety. When you reach the fork, the signpost is gone, but there are two people standing there. You know that they will answer only one question between them. What question can you ask either of them that will tell you which road is safe? (Grossworth et al., 1998, p. 81)

The mediator in conflict negotiations is similar to the person at the crossroads, the path she or he chooses must reflect the correct and necessary choices that the situation demands. Thus from the mediator's perspective you would ask either person: "Which road would the other tell me was safe?" and take the opposite. The liar would tell you the unsafe road, because the truth teller would point out the right road. The truth teller will say the truth - that liar would name the unsafe road. Each will point out the unsafe road, so the mediator would take the other one. (adapted from Grossworth et al., 1998, p. 118). If there is an overriding point to be taken from this is that despite what may seem like a deep engagement with the parties to a conflict resolution and finding a pathway towards the desired resolution between them, a mediator has an abiding interest to remain impartial to taking sides in the conflict itself. In some situations where strategic decision making is demanded, there is a constant balance between the irrelevance of neutrality and the deep commitment to resolution which is fraught with calims, counter-claims, negotiation *and* contingency. In this continuum the mediator is a participant in a similar sense to the ethnographer but has a duty to make a positive impact in the *lexis* of conflicting groups.

Strategies for coping with decision-making are mainly concerned with either making sure that an appropriate decision-making framework is in place, or intulogical thought processes are followed in arriving at a decision. However, people differ in their willingness or ability to engage with different levels and extents of cognitive processing. Cacioppo & Petty (as cited in Bouckenooghe at al., 2007, p.

608)) surmised that there were two psychological demands in decision-making: those that have a need for cognition (NFC) and those that have a need for cognitive closure (NFCL). The former category is characterized by enjoyment of thinking, and the latter the extent to which ambiguous responses are experienced aversively. The first category is likely to have a preference for observation, and the second to have a preference for order and structure to preserve past decisions and safeguard future knowledge. However, recently Neuberg, West, Judice and Thompson (as cited in Bouckenooghe et al., 2007, p. 608) have assumed that multi-dimensionality of cognitive decision-making styles is characterized along two orthogonal dimensions: the need for cognitive structure (NCS) and decisiveness (DEC). Psychologists are generally concerned with predicting behavioral and thought-processing types as much as they are with the process of decision-making themselves. This requires alteration in the framework of the thought processes around decision-making, rather than the concrete methodology of problem-solving in order to look at problems in inter-personal or workplace contexts.

Emotions and Decision-Making

As Chuang suggests, positive emotions may be advantageous for making choices and judgments, "positive emotional stimulation provides more activation and allows people to process more information effectively than do negative emotions" (p. 65). Others have argued that negative emotions tend to draw focus on situations requiring consideration and thus help decision-makers solve problems by pointing to the need to reduce stress, which motivates problem-solving (Chuang, 2007, pp. 63-64). However, the opposite has also been claimed: that when people experience positive emotions in problem-solving they are more likely to devote more attention to it, but when they experience negative emotions, they may divert attention from these by focusing on positive aspects of the self (Chuang, 2007, p. 65). People may make more benevolent judgments when they perceive faces or pictures as pleasant; hence positive mood does affect decision-making but not necessarily the accuracy of decision-making. According to Herrald and Tomaka (2002), pride may be associated with increased performance (problem-focused coping) and less co-support seeking than shame or anger. Positive emotional states may result in more problem-focused coping and less emotion-centered coping while negative emotional states involve more emotion-focused coping and less problem-centered coping. As Chuang (2007) states,

> Negative mood states facilitate systematic and analytic information processing and trigger more detail oriented, bottom-up, and vigilant processing styles. Positive mood states correspondingly tend to facilitate less systematic and analytic information processing and generate more top-down, schematic, and heuristic processing styles. (p. 66)

Generally, it can be assumed that positive emotions may help problem-solving by processing information efficiently. However, the constraints of the problem may inhibit this consideration due to the fact that people are often not required to act on their judgments. Information processing is often regarded as subservient to decision-making, so if knowledge is limited, people may restrict their decision-making due to not having all the available information (Chuang, 2007, p. 64).

Interestingly however, there is an element to decision-making from an information-processing perspective that is self-referential. According to the Emotional Congruency Model (Bower, 1981), emotions will create access to stimuli that resemble associated information. They will enhance the capacity for classification or will help to form "associative networks with valences consistent with their emotional state" (Chuang, 2007, p. 73). A distinction can also be made between task-related moods and ambient moods. The former are of shorter duration and have higher causal specificity than ambient moods, which are more pervasive and have lower cause specificity. If a negative task-related emotion is experienced, this may inform decision-makers that their current situation is problematic. This in turn may cause decision-makers to examine information more carefully, improve the accuracy of choices, and making more use of analytic strategies (Chuang, 2007, p. 73). As Martin and Delgado (2011) claim that the ability to control emotion is seen as a necessary part of adaptive functioning. Emotional regulation is a process of applying cognitive strategies which result in a change in affective experience, brought about by emotional stimuli (Martin & Delgado, 2011, p. 2596).

As Zhong (2011) suggests, emotion and deliberative thinking may produce different outcomes in decision-making along a moral dimension also. One of Zhong's finding is that deliberative thinking may decrease altruism. Some decision-making leads to moral dilemmas such as company layoffs. Managers in the position of making layoff decisions may experience ambivalence between the discomfort of letting someone go and doing the right thing for the company. Such decisions seem calculated but are consistent with collective interest (Zhong, 2011, p.16-17)

Group Decision-Making

Groups seek to work together to use a proven methodology to provide a right course of action for any given problem. As Steel, Regan, Colyvan and Burgman (2007, p. 352) point out, within standard decision-making the goal is to produce the maximum expected utility from any given situation. After probability values have been chosen (based on empirical evidence), and utility expectations defined, decisions are based on ranking various actions; the most positive outcome will be the one with the greatest expected utility. However, there is some doubt that 'right' may equate with objectively correct. However, as Steele, Regan, Colyvan and Burgman (2007, pp. 352-353) suggest, given the right conditions groups can arrive at objective decisions if a) group beliefs align with objective chances, and b) one option emerges that has the greatest expected utility according to all plausible probability. This is termed sensitivity analysis and attempts to account for the robustness of decision-making given small differences in probability and utility value (Steele at al., 2007, p. 353).

Decision-making by consensus in groups is often in response to intuitions about good sense decisions. Thus group views and opinions may be modified and updated along the way to forming an opinion. However, as Steel et al. (2007) suggest, convergence on a single solution may only be achieved in communication environments in which there is a communication of respect. The updating of group opinion is also paralleled by Bayesian methodologies in which the probability of an event is based on an entire set of knowledge, and opinion is updated in the light of new knowledge. However, sometimes group consensus will not be reached, group members may be alienated in such a process, or a compromise may be reached by negotiation. However, group decision-making is very useful in the contexts of complex decisions where a decision cannot be made within the knowledge frame of any one individual.

Differences in group leadership styles can also elicit differences in working behavior. Gonzalez and Tyler (2008, p. 448) claim that behavior exhibited when a leader is present or absent may change according to whether the leader is autocratic (work production drops when the leader is absent) or democratic (which production falls much less). Thus the inference is made that the participatory style of democratic leadership is inherently motivational. However, any workplace may be shaped by both internal and external motivations. Contingencies of the internal environment act as drivers for behavioral choices. The external environment may also influence motivations through determining the costs and benefits of various styles of behavior. Thirdly, there are motivational drivers intrinsic to the individual that develop from sources within the person and reflect their own desires, emotions, beliefs and behaviors. Gonzalez and Taylor (2008, p. 448) argue that when external contingencies are greater, individual differences in behavior are less readily observable, but in the absence of such external pressures, more individualistic behavior emerges that reflects people's beliefs and attitudes.

Workplace efficacy may also be modified by concepts of legitimacy. This inspires an associated belief that decisions made and enacted by others are in some way right or acceptable. This may make it easier for leaders seen as legitimately controlling workplace incentives or sanctions (Gonzalez & Tyler, 2008, p. 449). Furthermore, research also shows that people are more willing to accept decisions that they believe are arrived at in a fair way. Having a voice within group decision-making may have an interpersonal or expressive worth that is independent of the decision made. Often in group decision-making people may focus on the fairness of decision-making procedures and on interpersonal treatment, which leads to factors involved in the dynamics of decision-making (which are distinct from the outcomes of decision-making). As Gonzalez and Tyler (2008, p. 455) suggest, group membership is seen as important for status and feelings of self-worth. People in social environments may monitor these for determining the manner and extent to which they should engage in and make investments of self within the group. The result is a constant negotiation in identity formation (Gonzalez & Tyler, 2008, p. 455).

Strategic and Managerial Decision-Making

Strategic decision-making describes the application of common sense to dilemmas, problems and ambiguous situations. Common sense is defined by the qualities of soundness, practicality, validity, and may be subject to majority concurrence (Dinur, 2011, p. 696). Common sense in American parlance is also associated with concepts of average care, skill, and judgment (Dinur, 2011, p. 697). It is seen as the product of a 'reasonable' person's' thought processing, (not necessarily indicative of specialized knowledge or training), and it may be valid and applicable to a given dilemma or problem. Common sense decisions are derived from obtaining necessary information and having a theory or framework to support a systematic application of thought procedure to a set of issues or conflicting value inputs to arrive at a reasoned pathway to a course of action or best result. They thus involve having knowledge about the current environment (A), a desired future environment (B), and a pathway involving a formulated response that leads from A to B. Common sense can be distinguished from common knowledge, and also, under certain circumstances, from commonly accepted truth. The relationship of common sense to truth may be complex. What is accepted by the general public as the 'truth' at any given time may not necessarily correspond to factually accurate information; it may be based on hearsay or be influenced by political or 'folk-knowledge' beliefs (Dinur, 2011, p. 697). This implies that common sense not only differs between people, but is constructed socio-culturally.

The cognitive components of common sense include both logic and intuition. This combination has led Albrecht (2007) to coin the term "intulogical thinking"—the combination of intuition and logic. Intuition is partly associated with decision-making based on experience and feeling, but is also the result of decision-making in circumstances of time shortage and lack of other information. It is thus related both to the quality of decision-making and the use of expertise. The problem with intuitive thinking is its perceived riskiness – it is slightly paradoxical – leading either to devastating results or to optimal solutions (Dinur, 2011, p. 697). Intulogical thinking, on the other-hand combines both intuitive thinking and logical development. Furthermore, the intuitive dimension of common sense can be divided into the two factors of experience and causality. Experience informs intuition and common sense by adding to the repertoire of skills, knowledge, and memory traces, both in social and professional spheres, skills that the individual can draw on to derive an informed opinion or arrive at a decision. Causality relates to the intuitive (or even statistical probability) that two or more events are causally related. Causality is a major contributing cognitive component to reasoning by both induction and analogy. Thus experience informs intuition and vice versa, leading to the requisite knowledge framework to inform decision-making by inference, induction, analogy, intuition, and past experiences.

Decision-making is seen as a core activity of management practices, and is one of main activities that separate employers from employees. The contributing factors to decision-making in management include the support system (or decision framework), knowledge, and information availability. Within the contexts of timeframes and resourcing constraints, managers seek to make decisions that progress work activities smoothly from information base to knowledge and from project management to resource building. The information component of decision-making takes place within both decision and time frameworks. It is often characterized by being incomplete to complex. A major factor in decision-making is the reduction of ambiguity, which is understood as being in a state of incomplete knowledge. In contrast, equiviocality is understood as having several contradictory or competing frameworks for decision-making. Both ambiguity and equivocaility can make decision-making difficult. Figure 1. illustrates the relationship between ambiguity, equivocality, uncertainty, and complexity in decision-making (after Dinur, 2011, p. 698).

Figure 1. Dimensional Factors in Decision-Making (after Dinur, 2011, p. 698, Zack 2007)

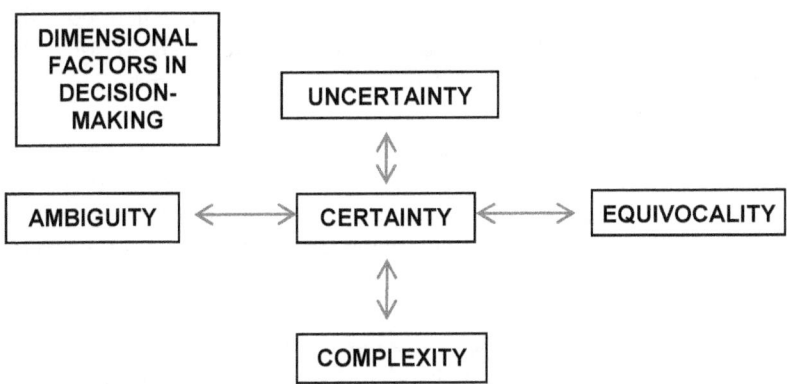

Managers typically tend to engage in two types of decision-making: structured and basic. The routine or commonplace decisions are described as structured because they are characterized by being defined by clear information channels and have certainty of cause and effect; in other words, they are procedural and predictable (Dinur, 2011, pp. 698-699). Often they are based on standard processing and past experience. Structured decisions take place within a common sense decision-making framework. One heuristic fault in this process is that it may lead to managers taking the path of least resistance when faced with task uncertainty (Dinur, 2011, p. 700). Basic decision-making, termed "thin-slicing decisions" (Gladwell, 2005) are characterized in comparison by being creative, innovative, complex, and often non-recurring. These have uncertainty in terms of cause and effect as well as unstructured decision-making criteria (Dinur, 2011, p. 699). Thin-sliced decisions involve heuristic 'rules of thumb' and the use of intuitive judgment.

However, in unstructured decision-making environments where there is either too little or too much information, it is possible to apply a variety of decision-making frameworks to minimize risk and arrive at a supportable decision. They are thus termed decisions necessitating 'uncommon sense.' The technique of thin-slicing involves the use of reduction to discern important factors regardless of available information or experience. The term 'analysis paralysis' is used to describe the state of indecision arising from information overload (Gladwell, 2005). Perceptual biases also lead to unstructured decision-making environments, which may themselves be the result of distorted inferences based on personal history or mass perception.

Dinur (2011) uses the term "managerial common sense" (MCS) to refer to the making of decisions under uncertainty where both the framework for analysis and the information required to solve the problem are present. If the latter is not present, the manager needs to be knowledgeable enough about the organization to obtain them. In order to exercise "managerial uncommon sense" (MUS), Dinur (2011) suggests that managers use an external focus of control or consult with outside expertise with distinctive competencies. When decisions are required in which managers do not have all the available information or a suitable decision framework, this necessitates outside or expert advice (which requires thinking outside the box). It may involve employing heuristics based on individualized experiences or rules of thumb. This highlights the point that the rationality exercised in common or uncommon sense may be an indeterminate concept, as it is difficult to measure differences between rational and irrational decision-making under 'uncommon sense' circumstances. Decisions derived from managerial uncommon sense may involve more risk, inviting failure or empowering transformation. Dinur (2011, p. 706) suggests that in situations involved in making decisions of low task certainty, managers resist using standard procedural frameworks.

Values and normative standards can also influence decision-making by either providing coherence to an individual's behavior or generating behaviors that influence group conformity. Poor decision-making may occur when information filtering conflicts with perceptions of individual values; consequently strategic values may affect their field of vision (Steptoe-Warren et al., 2011, p. 244). Within organizational communication, two kinds of values may operate: vision (why does a company exist?) and operating values (the ways processes are achieved while meeting company mission) (Steptoe-Warren et al., 2011, p. 245).

Strategic decisions are seen as the most important decisions that managers may make. A strategy is defined as a distinctive and predetermined pattern in an organization's decisions and activities, which leads to key strengths and developments which distinguishes the organization from other organizations. Strategic decisions can be divided into two types: content-based strategies and processes-based strategies.

Strategic decision is characterized by either a three or five-step process. The three-step process is described by, firstly, the characteristics of problem formulation and analysis of the objective setting the problem occurs in, secondly, the identification and generation of alternative solutions, and thirdly, the analysis and choice of a feasible alternative (Nooraie, 2008, p. 643). The features of the five-step process (after Fredrickson, 1984) are: firstly, situation diagnosis, secondly, alternatives generation, thirdly, alternatives evaluation, fourthly, selection of viable alternatives, and fifthly, integration of these within the operating business (Nooraie, 2008, p. 643). Furthermore, Nooraie (2008) argues that managerial decision-makers need to consider both the productive activity (efficacy) and the human contact (ethics) that arises from workplace decisions, which are judgments about the inherent aspects of the action chosen.

Within the theoretical framework of strategic decision-making, the nature of the decision made will have an influence on the nature of the process to be used. Similarly, the degree of the decision's magnitude of impact may determine the extent to which a manager may be more cautious and adhere to a rational decision-making process. If the impact of the decision is likely to be great, managers may act more conservatively and adopt more formal planning processes in their decision-making. However, it is doubtful whether a strategic decision can necessarily be termed high or low quality based on its outcome. As Nooraie (2008, p. 646) suggests, a good decision can lead to a bad outcome if its implementation is poor, and the collateral damage of a bad decision can be minimized with extraordinary capability in risk mitigation.

How well a decision is carried out may be determined by timeliness or speed in decision-making, the acceptability to interested others, and adaptiveness to change. Thus an effective implementation phase may be determined by how well the selected alternative is accomplished, how well the decision goals are achieved, and how well problems are solved (Nooraie, 2008, p. 646). Similarly, there is a correlation between the rationality of a decision-making process and the quality of the output. The choice of action in a strategic decision-making process may also be influenced by contextual variables, which can lead to adjustments in the manager's courses of action. Thus, contextual factors may have an indirect effect on the decision process output. The relationship between the decision magnitude of impact and the quality of the decision process can, according to Nooraie (2008, p. 651), be said to be mediated by the rationality of the decision-making process. Finally, the quality of the decision process output is likely to be better if the decision-making process is more rational.

Furthermore, strategic decision-making in medium and larger organizations is likely to be more rational than that in smaller-sized organizations due to two reasons: First, larger organizations have the necessary resources to conduct systematic research. Secondly, managers in larger organizations are more frequently paid employees (rather than owners), and thus may act as agents who need not be ultimately accountable, (a pressure that might otherwise constrain their actions and decision-making) (Nooraie, 2008). Furthermore Nooraie (2008, p. 651) suggests that the rationality of decision-making is not affected by ethnicity or gender, but more often by age. Younger managers who have less experience are more likely to make less rational decisions and to seek more support in their decision-making.

Strategic thinking is concerned with the changeability of organizational environments. Strategists need to be able to detect and act on changes affecting the organization both internally and externally. Internal factors include the culture of the organization, resources, and processes. External factors include technological impacts, environmental change, market-place changes, legislation, and politics (Steptoe-Warren et al., 2011, p. 239). Strategic planning thus includes two distinct processes – planning and thinking. Planning is concerned with analysis and creating formalized systems and procedures for organizational actions. Thinking is the domain within which this activity takes place and may involve synthesis, intuition, creativity, innovation, and analysis at many levels of the organization (Steptoe-Warren et al., 2011, p. 239). Such processes involve constant adaptation and improvement. They may include environmental scanning, visual representations of the organization's objectives, strategy mapping of organizational inter-relationships, competencies associated with problems or good performance, knowledge of enablers, and barriers to success (Steptoe-Warren et al., 2011, p. 240).

Strategists are frequently concerned with identifying and maximizing competencies, the skills required to perform particular activities. As Steptoe-Warren et al. (2011) suggest, "competency refers to the sum of experiences and knowledge, skills, traits, aspects of self-image and social role, values and attitudes" (p. 241). Competency is inherently defined by knowing what will (and what will not) work in a given circumstance or with a certain set of problem definitions. According to Liedtka (cited in Steptoe-Warren, 2011, p. 242), strategic decision-making competency has five characteristics:

1. Knowing how different sections of the organization influence one another and how they inter-relate with internal and external factors.
2. Creating a fit between existing resources and opportunities in the market-place.
3. Understanding the inter-connectivity and opportunities between the organization and the market-place, including competition.
4. Hypothesis testing – asking 'what if?' questions and managing risk.
5. Utilizing and maintaining an opportunistic viewpoint.

Michell, Shepherd and Sharfman (2010) argue that managers with greater metacognitive experience more frequently perform good decisions based on more stable, or at least, less erratic decisions. Strategic decisions are those important choices which may direct significant organizational resources and set precedent and direction for the organization. They may be influenced by a manager's prior knowledge and experiences, organizational context, and the dynamic nature of the organizational environment (Michell et al., 2010, p. 683). However, in more hostile environments manager's may make more erratic decisions, but for in those environments with high dynamism, the correlation between hostility and erratic decision-making is lower than for those experiencing the same hostility in an environment of lower dynamism (Michell et al., 2010, p. 683). Hostile environments may lead to disrupted information processing. Thus managers may need to be adept at information filtering, be able to ignore distracter interference, and avoid making decisions based on minimal data due to the influence of anxiety (Michell et al., 2010, p. 688). Uncertain environments may also negatively affect decision-making motivation.

According to Michell et al. (2010, p. 684) effective decisions are characterized by quality outcomes that exhibit reliability, adaptability, and performance, but erratic decisions can lead to lower organizational performance. Perceptions of the decision-making environment are inherently related to the reliability of the decision outcome. Erratic decision-making is often the result of inconsistent judgments on the part of the manager which affects the quality and direction of the organization. As Michell et al (2010, p. 685) suggest, judgment quality is a function of three factors: matching index (the appropriateness of the decision-making framework used according to the decision inputs); environmental predictability (the extent to which the use of an environmental model can explain environmental variance); and response consistency (the reliability of the execution of the decision). Michell et al. (2010) argue that metacognition or the conscious reflection on decision-making and the ability to understand and control it, is central to learning effectiveness in objective planning, adaptation, and implementation. Metacognition is thus understood as a form of quality control process and may preserve important knowledge.

As McKenzie, van Winkle and Grewal (2011) suggest, while decision-making is an intrinsic part of all business activities, "sound decisions rely on having the right knowledge in the right place at the right time" enabling effective action (p. 403). The context of field of experience for the use of knowledge in decision-making is important and varies from the level of surface knowledge for simple decisions, to intuition and insight-based creative decisions, to the tacit use of expertise (McKenzie et al., 2011, p. 403). Similarly the speed of decision-making has been linked to performance. Agility in dynamic environments is regarded as a core competency but relies on the capacity to learn and respond, to disregard irrelevant information, and to continuously be able to update and reconfigure new competencies, knowledge, and skills.

Consequently, the approach to decision-making is as important as the context, the sensitivity of the framing process, and the decision type. There are generally considered to be four types of decision contexts within the modern organization. First is the crisis or emergency decision-making context, which is concerned with the identity, purpose, or values of the organization. It may require expert judgment and the involvement of stakeholders at the highest level. It may also be characterized by 'triple-loop' systems of learning and unlearning, rapid feedback cycles, rapid courses of action and diagnosis, the use of expert decision-making, and pattern analysis. The second type of decision-making is strategic decision-making. This is understood as decision-making within a complex domain of emergence and it involves pattern detection, integrative thinking, insight, intuition, double-loop learning, and reflection. It is usually carried out by senior management and teams informed by experts. The third decision type is operation decision-making, characterized by complicated decisions in the domain of experts. It utilizes evidence-based research principles, understanding causality, single-loop and double-loop learning, and is usually carried out by management or supervisors. The fourth type of decision-making is tactical decision-making, which occurs in the domain of best practice. It is characterized by the use of descriptive information and explicit knowledge, know-how, and rules of thumb. It is generally carried out by practitioners, and involves awareness, memorizing, understanding, absorbing, explicit knowledge, and single-loop learning.

These decision types operate within a basic framework of simple, complicated and, complex decision-making. Simple decisions are those characterized by clear cause and effect linkages, have readily identifiable parameters and influences, and produce a foreseeable outcome. Complicated decisions require more sophisticated evaluation and expertise, and may involve the interaction of a multiplicity of variables. Complex decisions are 'grey' decisions which may have no right or wrong answers. They may be decisions which have large consequences and more continuing influences of cause and effect, with unpredictable outcomes. McKenzie et al. (2011, pp. 407-408) argue that each decision context requires a different weighting and ordering of decision-making dynamics. This includes intelligent sense-making to gauge the reality of the problem, identifying options in order to conceive a response, and making choices that lead to outcomes. Many decision-making environments require an emotional intelligence as well as the application of logic. Uncertain situations may lead to ambiguous and contradictory situations which heighten anxiety. In these situations, the emotional response is to reduce anxiety, but the logical response is to delay and live with tension to allow more time to understand the contradictions and connections that effect outcomes. Detached thinking is used to diffuse tension and to find solutions which creatively and coherently reconcile differences of view and opinion (McKenzie et al., 2011, pp. 407-408).

McKenzie et al. (2011, pp. 407-408) also identify 10 kinds of mental bias that may affect decision-making:

1) Escalation- which involves commitment to a losing course of action stemming from lack of adaptation to new knowledge. It is corrected by bringing in external perspectives to challenge assumptions.

2) Anchoring- which is giving disproportionate weight to the information first received, corrected by testing different scenarios.

3) Status quo- which is preferring alternatives that result in no change. It is corrected by involving both internal and external partners to stimulate learning.

4) Sunk-cost- refers to making choices that justify past actions and allow weak projects to continue. It is corrected by keeping the decision-making purpose clear, by using external evaluation, and with the use of decision-support technology.

5) Confirming evidence- which is seeking information that supports an individual point of view and rejecting evidence which does not. This is corrected by stimulating debate and dialogue to gain oversight.

6) Framing- which is making choices about how to position a problem in relation to fixed reference points; it may lead to precise answers to the wrong questions. It is corrected by involving more stakeholders to search for alternatives, using common sense and intuition, as well as analysis.

7) Over-confidence- which is a tendency to be more positive about forecasts or prognoses than they warrant. Consideration of the

opposite or careful calculation of probabilities may overcome this, as well as the use of collaborative decision-making support systems.

8) Over-prudence or risk aversion- which is a tendency to adjust estimates too conservatively, frequently preventing new ideas from progressing. Establishing parameters around the risks involved in problem solutions may enable choices to be rationally made between alternatives.

9) Recallability- involves being over-influenced by past dramatic events. A combination of inquisitiveness and skepticism may overcome this.

10) Preference for outsiders- which involves valuing knowledge from external sources more than internal sources.

Other biases may include unusual selection, which is focusing on unusual information at the expense of all information, and insensitivity to sample size, which is a failure to take into account a limitation in resources (Greenbank, 2011, pp. 252-256). While this may stimulate changes in group belief, it may also lead to the inappropriateness of framing problems.

Organizational Trust

Trust involves give and take. Hoe suggests that there are three kinds of trust that refer to people management within organizations. They are strategic trust, organizational trust, and interpersonal trust (Hoe, 2007, p. 151). Strategic trust is concerned with leadership trust, organizational trust refers to the general level of trust that people have for the organization, its people, systems and processes. Interpersonal trust is not only the trust that employees have for their mangers but also for each other in the workplace. Almost universally, any definition of trust involves the willingness or tolerance of vulnerability under "conditions of risk and interdependence" (Hoe, 2007, p. 151). Trust thus involves positive expectations that interdependence will not expose either part to inordinate risk or vulnerability. It also involves the notion of reciprocality – that actions performed will be mutually beneficial or at least not mutually detrimental. It is also

Figure 2. Relationship between the type and characteristics of the decision and the kinds of mental bias which may be encountered in decision-making (after McKenzie et al., 2011, Pp. 404-409)

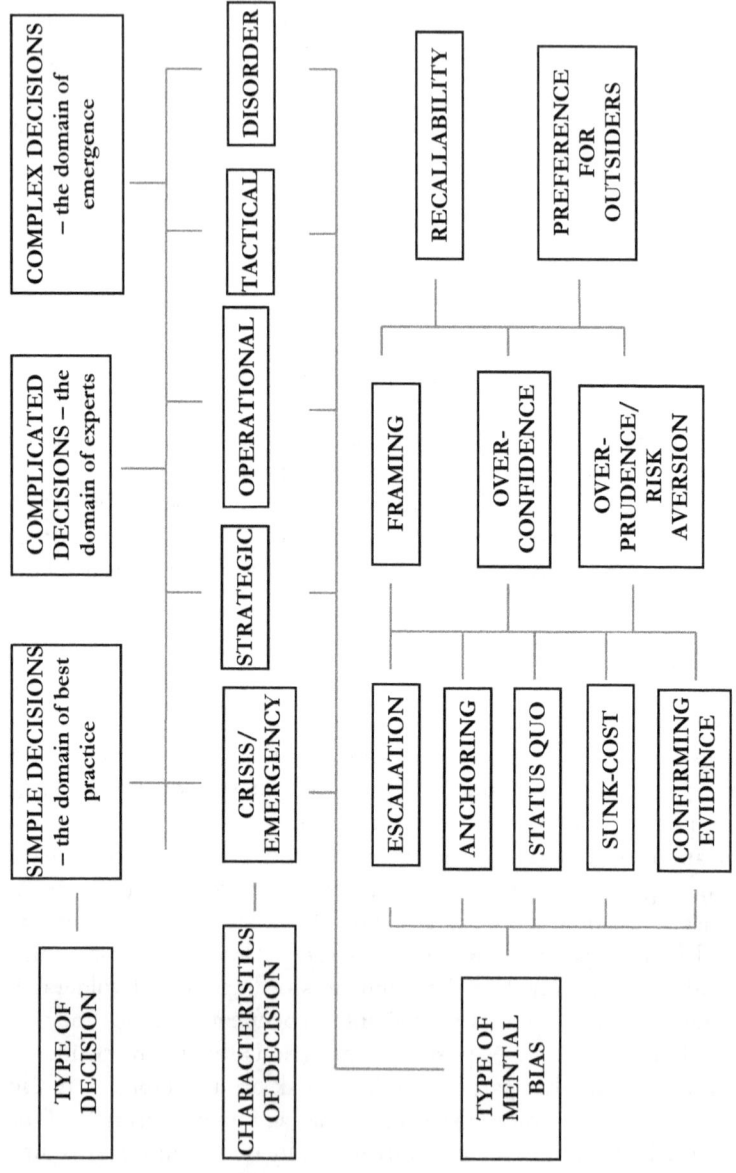

expressed in the notion that either party to the trusting agreement will not act opportunistically. In situations of uncertainty and personal conflict, trust is thus paramount. Trust can only develop when either party is willing to depend on the other party and to take risks. Trust thus involves investing a willingness for mutual interdependence in the other person, as well as a willingness to act in beneficial ways to the other person in conditions of uncertainty. Trust is essential to organizations in so much as it requires confidence and commitment in the acquisition and dissemination of knowledge. Knowledge acquisition and change needs to be filtered for possible beneficial or detrimental effects. The higher the level of interpersonal or organizational trust, the less 'checking' that needs to be done with the knowledge. It is based on repeated actions over time between the trustor and the trustee (Hoe, 2007, p. 153). Cooperation is derived from high personal trust. Trust may work through both formal and informal channels within an organization.

Economic and Commercial Trust

According to Heavey, Halliday, Gilbert and Murphy (2011, p. 2), the three key ingredients of trust in the business context are ability, benevolence, and integrity. As Ricketts (2001) states, "exchange requires trust" (p. 18). Even the simplest form of economic exchange – the barter—requires the necessary environment of honest relationships regarding the transaction. From whence does trust arrive? Is it intrinsic to the parties and process of the transaction? Or is it a factor external to the market, produced elsewhere as the result of activities in other spheres, such as education or religion, and then consumed or even subverted in transacting in the market place (Heavey et al., 2001, p. 18)? Trust is thus a desirable quality for market transaction as it encourages the conditions under which transactions can be made easily and efficiently. Consequently, the more trust that exists in the conditions for transaction, the more the costs of policing trust become economic.

Ricketts (2001) argues that the way organizations can be structured implies that trust is both a 'scarce and valuable' intangible quality. It is perhaps ironic that trust itself cannot be traded, though it can be acquired through association. As Ricketts (2001) states, "it is a quality which adhered to a particular person or organization as a result of their past history or their internal structure and cannot be transferred" (p. 18). Furthermore, Ricketts points out that both law and ethics (government regulation of markets or organizations) substitutes for private issues and perceptions of trust. Thus there is often a tension between interpersonal trust, organizational trust, ethical fair-dealing, and regulation.

If trust is moderate or high, usually lower transaction costs result. There are two problems associated with transaction costs (Ricketts, 2001, p. 19). Firstly, the problem of 'moral hazard' describes the way in which surveillance may preserve corporate responsibility. Secondly, the problem of adverse selection describes the situation in which people may pretend to offer higher quality services than they are able to deliver if 'quality' can't be assessed at low cost.

There is also another set of exigent transactional circumstances under which conditions of dependency in trust may be negotiated. This is referred to as 'hold up' or 'ex post contractual opportunism' (Ricketts, 2001, p. 19). A contractual relationship may entail dependency in which one group will entice another into making expensive investments of time, and then after resources have been committed, attempt to renegotiate the contract. Consequently both tangible and intangible resources (capital and knowledge and skills) are frequently more valuable in as much as they are transaction specific. Ex post contractual opportunism may be inspired by the ability of individuals and organizations to switch skills and resources at low cost to serve other groups. Consequently dependency tolerance is itself an intangible asset that is an economic characteristic of trust (Ricketts, 2001, p. 19). Economic trust is then related to property rights; where contractual terms may be complex, parties bargain to maximize their advantage. As Ricketts (2001) point out, vertical integration works best when there is a contract between parties for the use of specialized equipment, management, and labor inputs (when labor requirements are non-specific). However, when labor requirements are more specialized, trade across markets rather than integration may be preferable, thus preserving asset independence (Ricketts, 2001, p. 20). A further trust factor may be modified by the profit principle. Situations in which consumers may have a weak position for understanding their own interests and seek good advice are reinforced by the assurance that this advice is not affected by the interests of the provider. Consequently as Ricketts (2001, p. 21) points out 'low-powered' incentive may sometimes be more effective than 'high-powered' incentives.

Interpersonal Trust

If much of organizational communication is about reducing complexity and managing expectations, then as Gilbert (2005) suggests, trust may be about "skilled intervention by professionals" (p. 569). It may be guided by personal qualities, education, and by applying professional codes of behavior. Consequently 'governmentality' and trust are associated, functioning to reduce complexity. At the same time professionals and those who work in regulated workplaces need to be aware of the political nature of the systems in which they interact. Impersonal trust is based on an ideal or concept that pre-exists in the individual and is implicit in the professional systems that characterize the operational environment of professionals. However, it is not a good idea for professionals to always rely on these systems at the expense of their own judgment.

Trust, mistrust, and distrust are not necessary points on a continuum or opposites; rather they may be held in combination with one another at any one time. However, mistrust is motivational to some degree as it may imply that trust is possible. Thus it may enable the individual to manage change and diversity resulting from human unpredictability, or further to self-manage workplace situations through feelings of implied insecurity (Gilbert, 2005, p. 569). However, the trust dichotomy may be ruptured by abuse and also partially repaired by hope. However, while motivational, hope may in economic terms be interpreted as irrational due to its powerlessness (2005, p. 569). Conversely, although trust may frustrate hope through the management of expectations in professional activities that are constrained by material circumstances, hope on the other hand is a disposition of belief in a better future which is not so constrained.

Trust is thus a large variable in organizational learning. It is involved in both the knowledge acquisition and dissemination process. Interpersonal trust refers not only to the employee's "positive expectations about the supervisor's intentions" (Hoe, 2007, pp. 150-151) in difficult situations, but also to the level of compatible and genuine engagement with others in the work place within a framework of equitable disposition.

Consequently, trust is rarely a matter of the simple perception that one either holds it or doesn't. Trust is an enabler of successfully interpersonal and organizational behaviors but has different meanings for different people. As Herselman (2003) states, social trust is associated with the "establishment of social capital and co-operation" (p. 143) and does not necessarily correlate to trust in an organization or government. It thus has a psychological influence on peoples' attitudes, ideas, and emotions. Furthermore, trust is far harder to establish than it is to remove. Rarely is trust a matter of intelligence. Rather it is one of experience and involves "reciprocity and mutual expectations" (Herselman, 2003, pp. 143-144). Herselman (2003) describes trust as fragile, taking a long time to develop, capable of being easily lost, and re-established with difficulty. Distrust may be understood as a form of self-protection; it differs from mistrust in so far as the latter is characterized and active lack of trust and having no confidence in the behavior of others (Herselman, 2003, p. 144).

Organizational trust

Some forms of organizational trust may be due to different sociocultural factors rather than problems inherent in organizational culture itself. This is due to the fact that values and beliefs are frequently enculturated within individuals from an early age prior to the workplace setting (Herselman, 2003, p. 144). Recognition is a factor of workplaces that may improve self-confidence, reduce negative dispositions, and lessen the extent of wasted energy that people spend protecting themselves (Herselman, 2003, p. 147). Trust is thus a key source of social capital in organizations. Social capital itself inspires interdependence and allows co-workers to function cohesively (Herselman, 2003, p. 147). It is an important lubricant of a social system. Conversely, if employers or co-workers in positions of authority are seen to violate a psychological contract, trust may be reduced (Ohe & Martins, 2010, p. 1). As Ohe and Martins (2010, p. 4) and Macauley and Cook (2011) point out, trust in organizations is created by both managerial and personality factors. These are shown in the table below:

Personality factors involving trust	Managerial factors involving trust
Agreeableness	Information sharing
Conscientiousness	Work support
Resourcefulness	Credibility/sincerity
Emotional stability	Team management
Extraversion/introversion	Handling of conflict
Vulnerability	Empowering staff
Integrity	Feedback, praise, recognition
Competence	Coaching & development
Care/altruism	Transparency
Willingness to listen	Willingness to listen
Freedom to express opinions	Mediation and arbitration

Upwards, downwards, and horizontal communication and trade between supply chains is made easier if there is a useful level of organizational trust. From a public relations perspective, trust enhances confidence in organizations and makes it easier for them to do business. The fallout from lack of trust both internally and externally to an organization is registered both as a negative effect on reputation and in terms of the financial impact (Molloy, 2010, p. 55). Fahr and Irlenbusch (2008) describe trust as predominantly a property of individuals rather than organizations per se, providing their description of trust (inspired by individuals) as an ability to act as "boundary-spanning agents" (p. 469). They offer the following definition of trust between organizations: "the willingness of representative A to make herself [sic] and other members of her [sic] organization vulnerable to the actions of another representative B based on the expectation that the representative B will perform a particular action important to the members of A's organization" (Fahr & Irlenbusch,

2008, p. 470). This definition is seen to apply regardless of A's ability to monitor or control B.

Walls, Pidgeon, Weyman & Horlick-Jones (2004) highlight the role that trust has in mitigating risk perception in organizations. They define the following points in which it operates (Walls et al., 2004, p. 134):

- Focuses lay public concerns and responses to risk messages
- Contributes to the unacceptability of activities perceived as risky
- Stimulates social and political actions to reduce risks
- Leads to monitoring and the questioning of decisions taken by risk regulators and authorities
- Promotes the selective use of information sources.

Trust enhances commitment, which Heavey et al. (2011) suggest promotes cohesion and efficiency. Commitment is defined as a force which adheres an individual to a particular course of action. Trust and commitment are also motivational. According to Heavey et al. (2011, p. 3) the motivational element of trust is characterized by three factors. First by direction (what a person is trying to do), secondly by effort (how hard a person is trying), and lastly by persistence (how long a person may be engaged in a course of action). Furthermore, Heavey et al. (2011, p. 9) present the following equation for the estimation of workplace performance in relation to trust: Performance = function (aptitude X training X experience X motivation). All of the qualitative factors in this performance equation may involve elements of organizational trust.

Smith (2010, p. 19) suggests that for groups working collaboratively online, more clearly structured situations have clear processes, which produce outcomes with low ambiguity. Smith (2010, p. 20) adds that unconscious trust issues which have little to do with actual trust between members may influence the group. There are two main practical issues which may face group members: trusting their own ability to accomplish the task and trusting the knowledge of their peers (Smith, 2010, p. 21). However, Smith implies that online trust is a form of paradox; the ability to trust depends on the pre-existence of trust.

Mason and Lefrere (2003, p. 259) suggest that trust and collaboration are enablers of both organizational transformation and knowledge management. There are four main factors of facilitating online group work. Networking involves exchanging information among group members. It is characterized by coordination (which involves exchanging information among team members to position a group to achieve a common goal), cooperation (which involves exchanging information for mutual benefit and a common purpose); and collaboration (enhancing the capacity of another to meet a mutual goal) (Mason & Lefrere, 2003, p. 262). Accountability may enhance but can paradoxically undermine trust. In low-trust organizations with high accountability, transparency may remove secrecy but also may limit deception and misinformation (Mason & Lefrere, 2003, p. 262). Consequently, trust is seen as an important factor of economic intangibility. As both information and knowledge are economic goods it stands to reason that it is a performance enhancer and that high trust organizations should be valued (Mason & Lefrere, 2003, p. 265). Trust is thus predicated on consistency – consistency leads to predictability and conversely inconsistency leads to lower trust environments.

e-Commerce has two components – tangible components such as technology, vendors, and customers and intangible components such as the background or culture of each customer and the organizational culture in which they are situated (Gefen, Benbasat and Pavlou 2008, p. 103). Some uncertainty is needed in order for trust to be established. Trust may consist of two stages, an exploratory stage and a commitment stage. During the exploratory stage, interaction is based on disposition to trust and the background valency of organizational trust. The commitment stage of trust is characterized by the meeting or otherwise of expectations. Gefen et al. (2008) also suggest there are four stages of trust in information content. First is privacy, which describes a vendor's degree of openness. The second is that products and services should be transparent. Thirdly, the organizational factor should reflect a history, standards, obligation, and goodwill. The fourth is security, which applies to the relative completeness or understandability of information as it related to risks and guarantees (Mason & Lefrere, 2003, p. 106).

Organizational trust requires both structural assurance and situational normality (Mason & Lefrere, 2003, p. 109). The four main factors of overall organizational trust are an assessment of competence, benevolence, integrity, and predictability (Mason & Lefrere, 2003, p. 110). Calculus based trust refers to the business interaction defined by the customer-vendor relationship. Knowledge-based trust refers to the reliability and predictability of information exchanged between parties. Relational trust defines the overall ambit of the interaction between parties. For any given transaction, a customer will assess whether a vendor's website requires too much time, too much effort, or too much money (Gefen et al., 2008, p. 278). Gefen et al., (2008) points out that in contrast to traditional environments (face-to-face), consumers in online environments are not passive recipients of information. They are active participants who make assessments of vendor integrity based on information without an interpersonal component.

Two further factors influencing online trustworthiness include word of mouth (WOM) quality and perceived ease of use of the vendor's website (PEOU) (Awad & Ragowsky, 2008, 103). Benamati, Servaand Fuller (2008) state that in business-to-consumer electronic commerce (B2C), "trust without distrust results in an overconfident investor who may be exposed to financial risks; distrust without trust results in an overprotective consumer who is unwilling to realise [sic] the benefits of online banking" (p. 330). Furthermore, Purdue (2001, p. 2214) separates competence trust and goodwill trust. The former refers to the trust invested in the person or organization to accomplish the task and the latter refers to the emotional commitment with the other not to exploit vulnerability. Lim, Long Sia, Lee & Benbasat (2006, p. 241) suggest that there are three ways of measuring the intention to trust: through a willingness to provide personal information, through a willingness to engage in a purchase, and through a willingness to act on information.

Conclusion

Competency may be related to managerial cognition. This refers to the capacity to both attend to analytic detail and to be able to cut through to define salient features by use of strategic thinking strategies. Steptoe-Warren et al. (2011) characterize this as an ability to switch between habits of the mind and active thinking. Problem evaluation similarly involves two processes: an automatic pre-conscious process and deeper processes requiring more conscious cognition (Steptoe-Warren et al., 2011, p. 243). Thus, strategic thinking within an organization is based on knowledge of its past direction, where it is currently situated, and how it can adapt and survive in a dynamic future environment (Steptoe-Warren, 2011, p. 246). As Mele (2010) points out, decision-making can frequently be seen as a purely functionally-driven process characterized by rationality and the maximization of outcomes. However, this has been critiqued for its lack of humanistic realism and for the possibility that such functionalist-rationalizing behavior will not produce socially responsible decisions (Mele, 2010, p. 638). Instead, it is viewed as sensible to consider decision-making in terms of "bounded rationality" according to determined criteria and choosing among minimally satisfactory alternatives (Mele, 2010, p. 638). Organizational trust is an overriding factor in business efficiency and a key determinant of the interoperability of both inter-personal and web-based business functioning.

References

Albrecht, K. (2007). *Practical intelligence: The art and science of common sense.* San Francisco: Jossey Bass.

Awad, N. F. & Ragowsky, A. (2008). 'Establishing trust in electronic commerce through online word of mouth: An examination across genders'. *Journal of Management Information Systems, 24* (4), 101-121.

Benamati, J., Serva, M. A., Fuller, M. A. (2008). 'The productive tension of trust and distrust: The coexistence and relative role of trust and distrust in online banking'. *Journal of Organizational Computing and Electronic Commerce, 20,* 328-356.

Bouckenooghe, D., Vanderheyden, K., Mestdagh, S., Van Laethem, S. (2007). 'Cognitive motivation correlates of coping style in decisional conflict'. *The Journal of Psychology, 141* (6), 605-625.

Bower, G. H. (1981). 'Mood and memory'. *American Psychologist, 36,* 129-148.

Chuang, S-C. (2007). 'Sadder but wiser or happier and smarter? A demonstration of judgment and decision making'. *The Journal of Psychology, 141* (1), 63-76.

D'Ambrosio, B. (n.d.). 'Bayesian methods for collaborative decision-making'. *Robust Solutions.* Retrieved on 20 October 2012 from: http://www.robustdecisions.com/bayesianmethoddecisions.pdf

Dinur, A. R. (2011). 'Common and un-common sense in managerial decision-making under task uncertainty'. *Management Decision, 49* (5), 694-709.

Elliot, C. (1991). 'Competence as accountability'. *Journal of Clinical Ethics, 2* (3), 167-171.

Fahr, R., & Irelenbusch, B. (2008). 'Identifying personality traits to enhance trust between organizations: An experimental approach'. *Managerial and Decision Economics, 29*, 469-487.

Fredrickson, J.W. (1984). 'The comprehensiveness of strategic decision processes: extensions, observations, future directions', *Academy of Management Journal, 27*, 445-66.

Frost, R. (1920). 'The Road Not Taken'. *Poetry Foundation.org*. Retrieved from: http://www.poetryfoundation.org/ poem/173536

Gefen, D., Benbasat, I, Pavlou, P. A. (2008). 'A research agenda for trust in online environments'. *Journal of Management Information Systems, 24*(4), 275-286.

Gilbert, T. P. (2005). 'Interpersonal trust and professional authority: exploring the dynamics'. *Journal of Advanced Nursing, 49* (6), 568-577.

Gladwell, M. (2005). *Blink: The power of thinking without thinking*. New York: Back Bay Books.

Gonzalez, C. M., & Tyler, T. R. (2008). 'The psychology of enfranchisement: Engaging and fostering inclusion of members through voting and decision-making procedures'. *Journal of Social Issues, 64* (3), 447-466.

Greenbank, P. (2011). 'Improving the process of career decision-making: An action research approach'. *Education & Training, 53* (4), 252-266.

Grossworth, M, Salny, A. F., Stillson, A. (1998). *Match Wits with Mensa*. Cambridge, MA: Da Capo Press.

Heavey, C., Halliday, S. V., Gilbert, D., Murphy, E. (2011). 'Enhancing performance: Bringing trust, commitment and motivation together in organizations'. *Journal of General Management, 36* (3), 1-18.

Herrald, M. M., & Tomaka, J. (2002). 'Patterns of emotion-specific appraisal, coping, and cardiovascular reactivity during an ongoing emotional episode'. *Journal of Personality and Social Psychology, 71*, 245-261.

Herselman, S. (2003). "A little bit of distrust': causes and consequences of the trust-gap for work performance and relationships in a whole-sale company', *Anthropology Southern Africa, 26* (3/4), 143-149.

Hoe, S. L. (2007). 'Is interpersonal trust a necessary condition for organisational learning?'. *Journal of Organisational Transformation and Social Change, 4* (2), 149-156.

Janis, I. L., & Mann, L. (1977). *Decision Making: A Psychological Analysis of Conflict, Choice and Commitment.* Free Press: New York.

Lim, K. H., Ling Sia, C., Lee, M. K. O., Benbasat, I. (2006). 'Do I trust you online, and if so, will I buy? An empirical study of two trust-building strategies'. *Journal of Management Information Systems, 23*(2), 233-266.

Martin, L. N. & Delgado, M. R. (2011). 'The influence of emotion regulation on decision-making under risk'. *Journal of Cognitive Neuroscience, 23* (9), 2569-2581.

McKenzie, J., van Winkelen, C., Grewal, S. (2011). 'Developing organizational decision-making capability: A knowledge manager's guide'. *Journal of Knowledge Management, 15* (3), 403-421.

Mackenzie, R. & Watts, J. (2011). 'Can Clinicians and Carers Make Valid Decisions About Others' Decision-making Capacities Unless Tests of Decision-making Competence and Capacity Include Emotionality and Neurodiversity?' *Tizard Learning Disability Review, 16* (3), pp. 43-51.

Macauley, S. & Cook, S. (2011). 'Building trust into your organization'. Retrieved from: http://www.trainingjournal.com/feature/articles-features-building-trust-into-your-organization/

Mason, J & Lefrere, P. (2003). 'Trust, collaboration, e-learning and organizational transformation'. *International Journal of Training and Development, 7*(4), 259-270.

Mele, D. (2010). 'Practical wisdom in managerial decision-making'. *Journal of Management Development, 29* (7/8), 637-645.

Michell, J. R., Shepherd, D. A. & Sharfman, M. P. (2010). 'Erratic strategic decisions: When and why managers are inconsistent in strategic decision-making'. *Strategic Management Journal, 32*, 683-704.

Molloy, A. (2010). 'Trust but check'. *Chartered Accountants Journal, 89* (7), 55.

Owen, G. S., Freyenhagen, F., Richardson, G. & Hotopf, M. (2009). 'Mental capacity and decisional autonomy: An interdisciplinary challenge'. *Inquiry, 52* (1), 79-107.

Nooraie, M. (2008). 'Decision magnitude of impact and strategic decision-making process output'. *Management Decision, 46* (4), 640-655.

Ohe, H Von der, & Martins, N. (2010). 'Exploring trust relationships during times of change'. *Resource Management/SA Tydskrif vir Menslikehulpbronbestuur, 8*(1), 1-9.

Purdue, D. (2001). 'Neighbourhood governance: Leadership, trust and social capital'. *Urban Studies, 38* (12), 2211-2224.

Ricketts, M. (2001). 'Trust and economic organisation'. *Economic Affairs, 21* (2), 18-22.

Smith, R. O. (2010). 'Trust in online collaborative groups: A constructivist psychodynamic view'. *Adult Learning,* 19-23.

Steele, K, Regan, H. M., Colyvan, M. & Burgman, M. A. (2007). 'Right decisions or happy decision-makers?' *Social Epistemology, 21* (4), 349-368.

Steptoe-Warren, G., Howat, D. & Hume, I. (2011). 'Strategic thinking and decision-making: Literature review'. *Journal of Strategy and Management, 4* (3), 238-250.

Walls, J., Pidgeon, N., Weyman, A., Horlick-Jones, T. (2004). 'Critical trust: understanding lay perceptions of health and safety risk regulation'. *Health, Risk & Society,* 6: 2, 133-150.

Yang, S-Y. (2011). 'Wisdom displayed through leadership: Exploring leadership-related wisdom'. *The Leadership Quarterly, 22,* 616-632.

Zack, M.H. (2007). 'The role of decision support systems in an indeterminate world', *Decision Support Systems,* 43 (4), 1664.

Zhong, C-B. (2011). 'The ethical dangers of deliberative decision-making'. *Administrative Science Quarterly, 56,* 1-25.

CONCLUSION

The Organization and
Environmental Communication

As Van Woerkum and Aarts (2008) suggest, organizational survival may be predicated on an ability to reflect continually on the environment in a 'biotope' (p. 197). This is a categorization of the external or outside world according to the continuous reflection peculiar to the organization. This leads to the notion of a communication imprint (Van Woerkum & Aarts, 2008, p. 197). It is well known that communication is a fundamental percept of the strategic operation of an organization; however, there is a great variety of different organizational structures in communication with the outside world, each of which involves the notion of the organization tuning in to its environment (Van Woerkum & Aarts, 2008, p. 197). Such attunement to the environment, natural and man-made, exists both prior to and after the organization. Reich (1998, p. 15) says that all corporations "are, after all, creations of law; they do not exist in a state of nature." In human terms the corporation is can be either a small, medium, or large entity of organization that shares a collective common goal or mission. To what extent is it possible to associate the corporation with the natural world? Many insect and mammalian species are organized in families or groups and frequently we might describe the activities of a human organization in zoomorphic terms, such a 'hive of activity' for example, or the reverse, ascribe anthropomorphic qualities to animals, such as fish 'swimming in schools'. The use of tools adds further complexity and dimension of to human and animal socities. Arguably it is this factor, and the possibilities for intensional states and recursive thinking, that distinguish human corporations for animal incorporeality. Van Woerkum and Aarts (2008) suggest there are five kinds of groups that may have association and connection with an organization's environment. These include enabling groups (such as government agencies that make lawful production possible), input groups (that provide assets for an organization, e.g. money, personnel, knowledge, material necessities), output groups (consumers and citizens), groups with comparable goals with whom the organization competes or collaborates, and lastly normative groups that may advocate opinions about the organization (NGOs, media, religious organizations, political parties) (Van Woerkum & Aarts, 2008, p. 198). From this it is apparent that as Jurin et al., suggest, "[c]ulture and communication are interdependent" (2010, p. 189).

Van Woerkum and Aarts (2008) claim that it is possible to distinguish four types of organizational communication. The first is characterized by a diversity of channels. The main distinction to make in diversity of channels communication is that there is a difference between face-to-face communication and virtually mediated communication, (between one and two sided communication and that which reaches the human senses). However, communication channels are predicated on the intention of the sender (Van Woerkum & Aarts, 2008, p. 199). The second type of organizational communication is premised on the aims of the communication. Three kinds of information communication campaigns are distinguished: the information campaign which aims to increase knowledge, the persuasion campaign which aims to change attitudes/beliefs, and the mobilization campaign which aims to change behavior (Rogers and Storey, 1987). These in turn produce a hierarchy of effects. The third kind of organizational communication is to regard it as a form of intention – a means of learning about the environment and as a device with which to position the organization in an environment (Van Woerkum & Aarts, 200, p. 200). As Brinck (2008) suggests,

> Intentional communication is the nonphysical and nonverbal, co-ordinated interaction between (typically) two agents relative to a distal object in common space, the primary goal of which is to establish principally visual joint attention. It is essentially triadic, and enables the agents to influence each other's behavior indirectly, yet consistently, by way of the attention. (p. 4)

As well as the exegesis of interpersonal communication, many mass communication campaigns may use intentional communication to position the organization within its perceived and intended environment. The environmental communication of organizations may also involve a typology the modalities of which may be mutually influential. These include exploring, informing, relating and negotiation (Van Woerkum and Aarts, 2008, p. 200). Communication exploring in groups means knowing what they think about and act upon in matters related to the organization. However, there is also a cultural component in environmental communication. As Jurin et al. suggest, "The interaction of various microcultures within the larger domain of macroculture determines an individual's cultural identity. Membership in one microculture usually influences characteristics, beliefs and values of membership in other microcultures." (2010, p. 191). The goals of exploring are the reduction of uncertainty and the improvement of the readability of the environment. Exploring is an information gathering and information-processing task and is exemplified by scholarly or market research. Exploring is related to informing; as Van Woerkum and Aarts (2008) suggest, "one can only inform effectively by exploring perspectives and communications predispositions of the public beforehand" (p. 205). The message and mediums may be chosen and adapted according to the receiver's interests and previous preferences, and may be conditioned by information seeking behavior which leads to adaptation of the organizational aims and goals (Van Woerkum & Aarts, 2008, p. 205). Exploring and informing in organizational communication are becoming more influenced by social and dynamic marketing activities, such as discussion within communication networks, interpretative communities, or communities of discourse than individual based variables such as demographic market research (Van Woerkum, 2008, p. 205). A trend that Van Woerkum and Aarts discern is individualization in message production as global populations become more 'readable' through ICT phenomena such as social media (2008, p. 200). This involves people giving detailed information about themselves in exchange for more detailed information about the subject. This lessens the need for business to define a target market group in that the process is more geared towards self-selection, but does not necessitate any direct correspondence between customer input in product design (Van Woerkum & Aarts, 2008, p. 205). Whilst 'customisation' is not a new phenomenon in business terms, the quantity of data available

to companies to establish there market profile and customer preferences has increased exponentially with the use of virtual mediums.
However, whilst the barriers between businesses, customers and
products are becoming more permeable in ways which are exciting,
they are troubling at the same time. Despite the advances of genetic
engineering, people are not products and it is important that our
sense of environmental communication acknowledges our fundamental roots in ecological rather than synthetic processes.

If one of the main environmental communication objectives of
any organization is to inform others about itself, then the quality of
this information is to some extent influenced by the transparency of
the organization, (the answering of questions in an open and comprehensive manner). Advertising aims to make a formal link between
a target audience and a product or service provided by an organization, but it is not in the manner of a causal link (more a 'casual' link)
given that no direct reciprocity is expected. Consequently, much of
corporate communication aims to stimulate contact in a relational
way. There are three main strands to informing – relating, negotiating and branding. Relating is part of the process of gaining control
in the organization through informal channels rather than fixed
agreements (Van Woerkum and Aarts, 2008, p. 200). Consequently
relating is about contact with other relevant people or groups both
within and outside the organization. This contact may be on a scale
between exploring and diminishing relations and is characterized by
networking (Van Woerkum and Aarts, 2008, p. 202). Aside from informing, the sub-text to relating is stability and trust. This is achieved
by an understanding of others so that their behavior is made more
predictable and through the development of trust. Trust involves
commitment without the expectation of reciprocation. It allows social relationships to be more easily organized, reduces the amount of
error-checking in communication, and reduces transaction costs
(Van Woerkum and Aarts, 2008, p. 202). Networking allows networks to stay in touch with both formal and informal communication channels. This permits pre-emptive management by staying
aware of with contexts of prevailing attitudes and beliefs in the organization. Networking also permits greater understanding in communication. Furthermore, relating to others is also a part of organizational reputation and branding. Organizations that are open and responsive to stakeholders are more likely to be considered credible,
while organizational brands are firmed by invitations to relationships
of information exchange.

Negotiating is also a part of relating and communicating within a business environment. Negotiation takes place in an environment where two or more parties communicate "different interests" but "mutual dependencies", which makes forming transactions or concession and return possible (Van Woerkum and Aarts, 2008, p. 204). The characteristics of negotiation may be far-ranging – from simple exchanges and bargains where two parties with divergent claims arrive at a compromise (distributive negotiation), to integrative negotiation in which individual perspectives are set aside for a common perspective or frame in which both interests are fulfilled (Van Woerkum and Aarts, 2008, p. 204). Negotiations are often about shared interests rather than contractual details, but may require constant reframing and a variety of input from both parties.

Branding involves the monitoring of relations. It has been claimed that organizations that are more loosely coupled with their environment may be more able to discern outside events than tightly coupled organizations (Orton and Weick, 1990). This may be because they are less exposed to blind-spots. Consequently, the organization asserts its brand by reducing uncertainty. This is achieved by a) systematically identifying relevant others, b) formulating key questions, c) gathering answers, d) developing a unified picture, and, e) interpreting and evaluating findings (Van Woerkum and Aarts, 2008, p. 201).

The overall aim of informing is to give a concerted and clear account of the organizational position and where it is heading. However, organizations are also constrained by the environments they exist in and an organizational 'image or campaign' is one among many in any given market-place. Thus, corporate silence is not seen as a useful option. Furthermore, when there are different representations of an organization, trustworthiness may be an issue. Van Woerkum and Aarts (2008) say that the notion that the organization is the active sender and that the stakeholder/participant is the passive receiver may be misplaced. It is questionable whether 'receivers' actually exist in the communication model. Rather, they think that individuals construct perceived meaning in a process of constant 'visualization' or sensory association (2008, p. 201). In this model, both senders and receivers are active and involved in co-production of organizational meaning.

Thus there are varied dependencies between an organization and the groups, members, corporates, stakeholders and individuals who interact with it. The preference is for the organization to be 'in-tune' with its operating environment, indeed many of the problems that organizations may face are due to the absence of the latter, through lack of knowledge, misaligned procedures, or because they lack the necessary 'tools' and resources in order to 'read' the environment and the place of the human activity within it. Consequently, environmental communication, which focuses on the functioning and complexity of the human individual within the corporate ecosystem, is a process of learning and 're-orientation' which involves the goal of a fundamental synchrony between people and their environments.

References

Brinck, I. (2008). The role of intersubjectivity for the development of intentional communication. In J. Zlatev, T. Racine. C. Sinha, & E. Itkonen (Eds.). *The shared mind: Perspectives on intersubjectivity* (pp. 115-140). Amsterdam: John Benjamins Publishing.

Jurin, R. R., Roush, D., Danter, J. (2010). *Environmental Communication. Second Edition: Skills and Principles for Natural Resource Managers, Scientists, and Engineers.* New York: Springer.

Orton, D. and Weick, K. E. (1990). 'Loosely coupled systems: A reconceptualization'. *Academy of Management Review, 15* (2), 203-223.

Reich, R. B. (1998). 'The new meaning of corporate social responsibility'. *California Management Review, 40* (2), 8-17.

Rogers, E. M. and Storey, J. D. (1987). Communication campaigns. In C. R. Berger & S. H. Chafee (Eds.). *Handbook of Communication Science* (pp. 817-846). Newbury Park, CA: Sage.

Van Woerkum, C., & Aarts, N. (2008). 'Staying connected. The communication between organizations and their environment'. *Corporate Communications: An International Journal. 13* (2), 197-211.

ACKNOWLEDGEMENTS

Several of the essays in this book had their origins as peer reviewed articles or conference presentations. The author is grateful for all permissions for use.